Advances in Experimental Medicine and Biology

Volume 1213

Advances in Experimental Medicine and Biology presents multidisciplinary and dynamic findings in the broad fields of experimental medicine and biology. The wide variety in topics it presents offers readers multiple perspectives on a variety of disciplines including neuroscience, microbiology, immunology, biochemistry, biomedical engineering and cancer research.
Advances in Experimental Medicine and Biology has been publishing exceptional works in the field for over 40 years, and is indexed in SCOPUS; Medline (PubMed); Journal Citation Reports/Science Edition; Science Citation Index Expanded (SciSearch) (Web of Science); EMBASE; BIOSIS, Reaxys; EMBiology; the Chemical Abstracts Service (CAS); and Pathway Studio. The series also provides scientists with up to date information on emerging topics and techniques.
2018 Impact Factor: 2.126.

More information about this series at http://www.springer.com/series/5584

Gobert Lee • Hiroshi Fujita
Editors

Deep Learning in Medical Image Analysis

Challenges and Applications

Springer

Editors
Gobert Lee
College of Sciences & Engineering
Flinders University
Adelaide, SA, Australia

Hiroshi Fujita
Faculty of Engineering
Gifu University
Gifu, Japan

ISSN 0065-2598 ISSN 2214-8019 (electronic)
Advances in Experimental Medicine and Biology
ISBN 978-3-030-33127-6 ISBN 978-3-030-33128-3 (eBook)
https://doi.org/10.1007/978-3-030-33128-3

This Springer imprint is published by the registered company Springer Nature Switzerland AG.
The registered company address is: Gewerbestrasse 11, 6330 Cham, Switzerland

Preface

Deep learning is at the leading edge of artificial intelligence (AI) and is developing rapidly. In recent years, it has played an increasingly important role in medical image analysis. Deep learning is a subfield of machine learning and is based on deep neural networks (DNNs)—neural networks with more than one hidden layer. Convolutional neural networks (CNNs) are a subclass of DNNs that are especially useful for image recognition and classification and have been attracting a lot of interest from industry, academia, and clinicians.

In 2017, we organized the first mini-symposium on "Emerging Methods in Medical Image Analysis" at the 39th International Conference of the IEEE Engineering in Medicine and Biology Society (EMBC) held in Jeju, Korea. The main objectives of the mini-symposium were to bring together researchers in the field to explore current issues in medical image analysis and to report the latest research directions. The symposium was met with great interest and enthusiasm from the audience and researchers in the field. In 2018, we organized the second mini-symposium on "Emerging Methods in Medical Image Analysis" at the 40th EMBC International Conference held in Honolulu, Hawaii. With the overwhelming support from the research community and presenters, the mini-symposium had grown to double the session length of the first mini-symposium. In this second symposium, deep learning was the most popular technique reported among the research presented. This was largely because CNNs can self-learn high-level representations of image contents and achieve good performance in medical imaging tasks without the need to construct handcrafted features or acquire field knowledge about the image data. Skills in this area are also highly sought after by both industry and academia. Due to the widespread interest in this topic, many people, including Springer Publishing, have asked us to publish an edited book based on the mini-symposium presentations. Subsequently, a meeting was held among the presenters in Honolulu after the symposium. The publication of a book was supported by the majority. We have also invited other lead researchers in the field to contribute to the book.

This book aims at providing overviews of the use of deep learning in medical image analysis; sampling the roles of medical image analysis in key clinical applications and emerging applications; giving implementation examples and insights of medical image analysis studies; highlighting issues and challenges encountered by researchers and clinicians; as well as surveying and discussing practical approaches in general and for specific applications.

We hope that this book will provide a helpful reference for academics, clinicians, and researchers, as well as a useful learning resource for young researchers and graduate students in AI, medical image analysis, biomedical engineering, and other related areas.

We would like to thank Merry Struber, Editor, for her dedicated assistance in getting this book project off the ground and Deepak Ravi, Production Editor, for his tireless efforts in getting this book through the stages of production. We thank all the contributors for their enthusiastic support of this book. This book would not have appeared if not for all their efforts. We are particularly grateful to them for accommodating waves of short deadlines that were sprung on them. Finally, we hope that readers will find this book interesting, insightful, and helpful.

Adelaide, SA, Australia Gobert Lee
Gifu, Japan Hiroshi Fujita

Contents

Part III Applications: Emerging Opportunities

Part I

Overview and Issues

Deep Learning in Medical Image Analysis

Heang-Ping Chan, Ravi K. Samala, Lubomir M. Hadjiiski, and Chuan Zhou

Abstract

Deep learning is the state-of-the-art machine learning approach. The success of deep learning in many pattern recognition applications has brought excitement and high expectations that deep learning, or artificial intelligence (AI), can bring revolutionary changes in health care. Early studies of deep learning applied to lesion detection or classification have reported superior performance compared to those by conventional techniques or even better than radiologists in some tasks. The potential of applying deep-learning-based medical image analysis to computer-aided diagnosis (CAD), thus providing decision support to clinicians and improving the accuracy and efficiency of various diagnostic and treatment processes, has spurred new research and development efforts in CAD. Despite the optimism in this new era of machine learning, the development and implementation of CAD or AI tools in clinical practice face many challenges. In this chapter, we will discuss some of these issues and efforts needed to develop robust deep-learning-based CAD tools and integrate these tools into the clinical workflow, thereby advancing towards the goal of providing reliable intelligent aids for patient care.

Keywords

Machine learning · Deep learning · Artificial intelligence · Computer-aided diagnosis · Medical imaging · Big data · Transfer learning · Validation · Quality assurance · Interpretable AI

Introduction

Medical imaging is an important diagnostic tool for various diseases. Roentgen discovered that X-rays could non-invasively look into the human body in 1895 and X-ray radiography became the first diagnostic imaging modality soon after. Since then many imaging modalities were invented, with computed tomography, ultrasound, magnetic resonance imaging, and positron emission tomography among the commonly used, and more and more complex imaging procedures have been developed. Image information plays a crucial role in decision making at many stages in the patient care process, including detection, characterization, staging, treatment response assessment, monitoring of disease recurrence, as well

H.-P. Chan (✉) · R. K. Samala · L. M. Hadjiiski
C. Zhou
Department of Radiology, University of Michigan, Ann Arbor, MI, USA
e-mail: chanhp@umich.edu

G. Lee, H. Fujita (eds.), *Deep Learning in Medical Image Analysis*, Advances in Experimental Medicine and Biology 1213,
https://doi.org/10.1007/978-3-030-33128-3_1

as guiding interventional procedures, surgeries, and radiation therapy. The number of images for a given patient case increases dramatically from a few two-dimensional (2D) images to hundreds with 3D imaging and thousands with 4D dynamic imaging. Application of multi-modality imaging further increases the amount of image data to be interpreted. The increasing workload makes it difficult for radiologists and physicians to maintain workflow efficiency while utilizing all the available imaging information to improve accuracy and patient care. With the advances in machine learning and computational techniques in recent years, developing effective and reliable computerized methods to assist radiologists and physicians in image analysis at various stages of disease diagnosis and management during the patient care process has been recognized as an important area of research in medical imaging.

The attempt of using computers to automatically analyze medical images emerged as early as the 1960s [1–4]. Several studies demonstrated the feasibility of applying computer to medical image analysis but the work did not attract much attention, probably because of the limited access to high quality digitized image data and computational resources. Doi et al. in the Kurt Rossmann Laboratory at the University of Chicago began systematic development of machine learning and image analysis techniques for medical images in the 1980s [5], with the goal to develop computer-aided diagnosis (CAD) as a second opinion to assist radiologists in image interpretation. Chan et al. developed a CAD system for detection of microcalcifications on mammograms [6] and conducted the first observer performance study [7] that demonstrated the effectiveness of CAD in improving breast radiologists' detection performance of microcalcifications. The first CAD commercial system was approved by the Food and Drug Administration (FDA) for use as a second opinion in screening mammography in 1998. CAD and computer-assisted image analysis have been a major area of research and development in medical imaging in the past few decades. CAD methods have been investigated for various applications including disease detection, characterization, staging, treatment response assessment, prognosis prediction, and risk assessment for various diseases and with various imaging modalities. The work in the CAD field has been steadily increasing as can be seen from the trend of publications in peer-reviewed journal articles found by literature search in the Web of Science (Fig. 1).

Although the research in CAD has been increasing, very few CAD systems are used routinely in the clinic. One of the major reasons may be that CAD tools developed with conventional machine learning methods may not have reached the high performance that can meet physicians' needs to improve both diagnostic accuracy and workflow efficiency. With the success of deep learning in many machine learning applications such as text and speech recognition, face recognition, autonomous vehicles, chess and Go game, in the past several years, there are high expectations that deep learning will bring breakthrough in CAD performance and widespread use of deep-learning-based CAD, or artificial intelligence (AI), to various tasks in the patient care process. The enthusiasm has spurred numerous studies and publications in CAD using deep learning. In this chapter, we will discuss some issues and challenges in the development of deep-learning-based CAD in medical imaging, as well as considerations needed for the future implementation of CAD in clinical use.

Deep Learning for Medical Image Analysis and CAD

CAD systems are developed with machine learning methods. Conventional machine learning approach to CAD in medical imaging used image analysis methods to recognize disease patterns and distinguish different classes of structures on images, e.g., normal or abnormal, malignant or benign. CAD developers design image processing and feature extraction techniques based on domain knowledge to represent the image characteristics that can distinguish the various states. The effectiveness of the feature descriptors often depends on the domain expertise of the CAD developers and the capability of the mathematical formulations or empirical image analysis techniques that are designed to translate the image

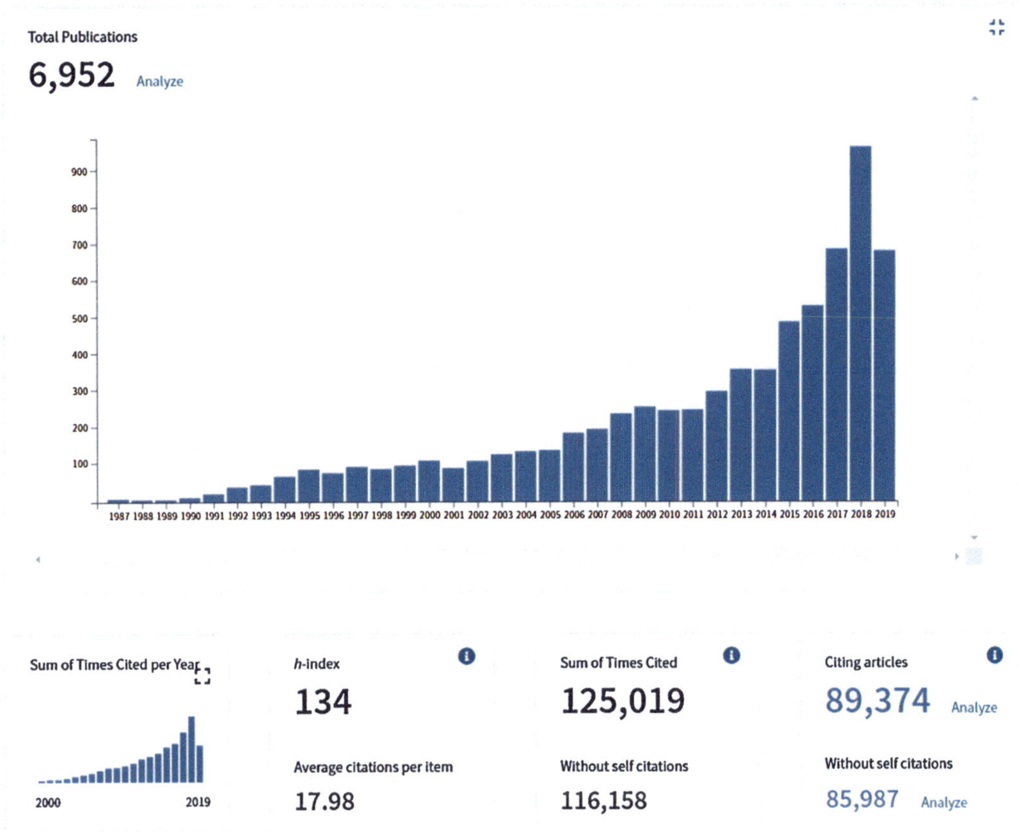

Fig. 1 Literature search for publications in peer-reviewed journals by Web of Science from 1900 to early July of 2019 using key words: ((imaging OR images) AND (medical OR diagnostic)) AND (machine learning OR deep learning OR neural network OR deep neural network OR convolutional neural network OR computer aid OR computer assist OR computer-aided diagnosis OR automated detection OR computerized detection OR computer-aided detection OR automated classification OR computerized classification OR decision support OR radiomic) NOT (pathology OR slide OR genomics OR molecule OR genetic OR cell OR protein OR review OR survey))

characteristics to numerical values. The extracted features are then used as input predictor variables to a classifier, and a predictive model is formed by adjusting the weights of the various features based on the statistical properties of a set of training samples to estimate the probability that an image belongs to one of the states. Conventional machine learning approach has limitations in that the human developer may not be able to translate the complex disease patterns into a finite number of feature descriptors even if they have seen a large number of cases from the patient population. The hand-engineered features may also have difficulty to be robust against the large variations of normal and abnormal patterns in the population. The performance of the developed CAD system is often limited in its discriminative power or generalizability, resulting in high false positive rate at high sensitivity or vice versa.

Deep learning has emerged as the state-of-the-art machine learning method in many applications. Deep learning is a type of representation learning method in which a complex multilayer neural network architecture learns representations of data automatically by transforming the input information into multiple levels of abstractions [8]. For pattern recognition tasks in images, deep convolutional neural networks (DCNN) are the most commonly used deep learning networks. With a sufficiently large training set, DCNN can learn to automatically extract relevant features from the training samples for a given task by iteratively adjusting its weights with backpropagation. DCNN therefore discovers feature representations through training and does not require manually designed features as input. If properly trained with a large training set that are representative of the population of interest, the DCNN features are expected to be superior to hand-engineered features in that they have high selectivity and invariance [8]. Importantly, since the learning process is automated, deep learning can easily analyze thousands or millions of cases that even human experts may not be able to see and memorize in their lifetime. Deep learning can therefore be more robust to the wide range of variations in features between different classes to be differentiated as long as the training set is large and diverse enough for it to analyze.

CNN can trace its origin to the neocognitron proposed by Fukushima et al. in the early 1980s [9]. LeCun first trained a CNN by backpropagation to classify patterns of handwritten digits in 1990 [10]. CNN was used in many applications such as object detection, character, and face recognition in the early 1990s. Lo et al. first introduced CNN to the analysis of medical images in 1993 and trained a CNN for lung nodule detection in chest radiographs [11, 12]. Chan et al. applied CNN to microcalcification detection [13, 14] on mammograms in the same year and to mass detection in the following year [15–18]. Zhang et al. applied a similar shift-invariant neural network for the detection of clusters of microcalcifications in 1994 [19]. Although these early CNNs were not very deep, the pattern recognition capability of CNN in medical images was demonstrated.

Deep CNN was enabled by several important neural network training techniques developed over the years, including layer-wise unsupervised pre-training followed by supervised fine-tuning [20–22], use of rectified linear unit (ReLU) [23, 24] as activation function in place of sigmoid-type activation functions, pooling to improve feature invariance and reduce dimensionality [25], dropout to reduce overfitting [26], and batch normalization [27] that further reduces the risk of internal covariate shift, vanishing gradient and overfitting, as well as increases training convergence speed. These techniques allow neural networks with more and more layers and containing millions of weights to be trained. In 2012, Krizhevsky et al. [28] proposed a CNN with five convolutional layers and 3 fully connected layers (named "AlexNet") containing over 60 million weights and achieved breakthrough performance in the ImageNet Large Scale Visual Recognition Challenge (ILSVRC) [29] that classified over 1000 classes of everyday objects on photographic images. AlexNet demonstrated the pattern recognition capability of the multiple layers of a deep structure. DCNNs with increasing depth were developed since AlexNet. He et al. [30] proposed residual learning and showed that a residual network (ResNet) with 110–152 layers could outperform several other DCNNs and won the ILSVRC in 2015. Sun et al. [31] showed that the learning capacity of a DCNN increased with depth but the capacity could be utilized only with sufficiently large training data.

The success of deep learning or AI in personal devices and social media, self-driving cars, chess, and Go game have raised unprecedented expectations of deep learning in medicine. Deep learning has been applied to many medical image analysis tasks for CAD [32–34]. The most common areas of CAD application using deep learning include classification of disease and normal patterns, classification of malignant and benign lesions, and prediction of high risk and low risk patterns of developing cancer in the future. Other applications included segmentation and classification of organs and tumors of different types, classification of changes in tumor size or texture for assessment of treatment response, or prediction of prognosis or recurrence. Because there are relatively large public data sets available for chest radiographs, thoracic CT, and mammograms, a large number of studies were conducted for lung diseases and breast cancer using the public data sets. Deep-learning-based image analysis has also been applied to fundus images or optical computed tomography for detection of eye diseases [35], or histopathological images for classification of cell types [36]. Most of the studies reported very promising results, further boosting the hype of deep-learning-based CAD. This new generation of CAD is called AI although these CAD tools still behave like a very complex mathematical model that memorizes information in its millions of weights and far from being "intelligent."

Challenges in Deep-Learning-Based CAD

CAD or AI is expected to be useful decision support tools in medicine in the near future. Other than detection and characterization of abnormalities, applications such as pre-screening and triaging, cancer staging, treatment response assessment, recurrence monitoring, and prognosis or survival prediction are being explored. Although no CAD systems with new AI techniques have been subjected to large scale clinical trials to date, experiences from CAD use in screening mammography may provide some insights into what may be expected of CAD tools in the clinic [37].

The conventional machine-learning-based CAD for detection of breast cancer in screening mammography is the only CAD application in widespread clinical use to date. These systems have been shown to have sensitivity comparable to or higher than that of radiologists, especially for microcalcifications, but they also mark a few false positives per case on average [38]. Although the performances of CAD systems are moderate, they may detect lesions of different characteristics than those by radiologists. The complementary detections by the radiologist and CAD can improve the overall sensitivity when radiologist reads with CAD. Studies have shown that radiologists' accuracy was improved significantly when reading with CAD [5]. CAD systems were therefore approved by FDA for use as a second opinion but not as a primary reader or pre-screener. Early clinical trials [39, 40] to compare single reading with CAD to double reading showed promising results. In the CADET II study by Gilbert et al. [39], they conducted a prospective randomized clinical trial at three sites in the United Kingdom. A total of over 28,000 patients were included. The screening mammograms of each patient were independently read in two arms; one was single reading with CAD and the other was their standard practice of double reading. The experiences of the single readers in the CAD arm were matched to those of the first readers' in the double reading arm. Arbitration was used in cases of recall due to the second reader or CAD. They found that arbitration was performed in 1.3% of the cases in single reading with CAD. The average sensitivity in the two arms were comparable at 87.2% and 87.7%, respectively. The recall rates at two centers were comparable in the two arms, 3.7% versus 3.6% and 2.7% versus 2.7%, respectively, but one of the centers had a significantly higher recall rate for single reading with CAD, 5.2% versus 3.8%. The overall recall rate therefore increased in the single reading with CAD from 3.4% to 3.9%. Gromet et al. [40] performed a retrospective review of the sensitivity and recall rate by single reading with CAD after CAD implementation in comparison to those of double reading before CAD use as historical control for the same group of nine radiologists in a single mammography facility. The first reading in their double reading protocol was also analyzed and treated as single reading without CAD. The study cohort contained over 110,000 screening examinations in each group. Arbitration by a third subspecialty radiologist was a part of their standard double reading protocol. A second radiologist was consulted for 2.1% of the cases interpreted by single reading with CAD but the consult might or might not be related to CAD marks. They reported that the sensitivity of single reading with CAD was 90.4%, higher than the sensitivities of either single reading alone (81.4%) or double reading (88.0%). The recall rate was 10.6% for single reading with CAD, slightly higher than the recall rate of single reading alone (10.2%) but lower than that of double reading (11.9%). These relatively well-controlled studies showed that single reading with CAD is potentially an alternative to double reading, with a gain in sensitivity but at the expense of increased recalls, which can be reduced by arbitration similar to that in double reading.

Table 1 Odds ratios (95% confidence interval) of increase in cancer detection rate and increase in recall rate obtained by comparison of single reading with CAD and double reading to single reading alone by Taylor et al. [41]

	Odds ratio of increase in cancer detection rate	Odds ratio of increase in recall rate
Single reading with CAD		
Matched ($N = 5$)	1.09 (0.92, 1.29)	1.12 (1.08, 1.17)
Unmatched ($N = 5$)	1.02 (0.93, 1.12)	1.10 (1.08, 1.12)
Double reading		
Unilateral ($N = 6$)	1.13 (1.06, 1.19)	1.31 (1.29, 1.33)
Mixed ($N = 3$)	1.07 (0.99, 1.15)	1.21 (1.19, 1.24)
Arbitration ($N = 8$)	1.08 (1.02, 1.15)	0.94 (0.92, 0.96)

N: the number of studies included in each group

Matched studies: the assessment before and after using CAD was on the same mammograms

Unmatched studies: the performance of mammography facilities after the introduction of CAD was compared to that before CAD implementation as historical controls. Different mammograms were interpreted in the two conditions

Taylor et al. [41] conducted a meta-analysis of clinical studies comparing single reading with CAD or double reading to single reading alone. They compared the cancer detection rate per 1000 women screened (CDR) and the recall rate, and estimated the average odds ratios weighted by sample size over the studies in each group (Table 1). The results showed that double reading with arbitration improved the CDR without increasing the recall rates. Single reading with CAD for the matched studies increased the CDR but with a wide variation; however, without the benefit of arbitration, the recall rate increased significantly. The increase in recall rate for double reading without arbitration was more than twice of that for single reading with CAD.

Taylor et al. revealed that there are large variations in the impact of CAD on the cancer detection rate ranging from 0% to 19%, and the recall rate ranging from 0% to 37%. Other than the differences in the study designs and radiologists' experiences in the studies, the variations may also be attributed to the varied ways that radiologists used CAD in the clinic. Some users might have misunderstood the limitations and performance of the CAD systems. They might have over-relied on the CAD marks and thus did not maintain their vigilance in searching for lesions while increasing their recalls. Others might have used CAD as a pre-screener or first reader to increase workflow. Although there were no systematic studies of how CAD was used in the clinic, Fenton et al. [42] noted that "radiologists with variable experience and expertise may use CAD in a nonstandardized idiosyncratic fashion," and "Some community radiologists, for example, may decide not to recall women because of the absence of CAD marks on otherwise suspicious lesions." Lehman et al. [43] compared reading digital mammograms with and without CAD by 271 radiologists in 66 facilities of the Breast Cancer Surveillance Consortium (BCSC). They reported that the average sensitivity decreased by 2.3% and the recall rate increased by 4.5% with the use of CAD. The decrease in sensitivity was a clear indication that the radiologists did not use CAD as a second opinion, which require the users to maintain their vigilance in interpretation and thus their sensitivity, but over-relied on the CAD marks for recall decisions. The authors acknowledged that "Prior reports have confirmed that not all cancers are marked by CAD and that cancers are overlooked more often if CAD fails to mark a visible lesion" and that "CAD might improve mammography performance when appropriate training is provided on how to use it to enhance performance."

The study by Cole et al. [38] demonstrated another facet of using CAD. They conducted an observer study to compare single reading with and without CAD using two commercial CAD systems applied to 300 screening cases (150 cancers and 150 benign or normal) from the Digital Mammographic Imaging Screening Trial (DMIST). All participating

Table 2 Observer performance study by Cole et al. [38] comparing single reading with and without CAD using two commercial CAD systems and 300 screening mammography cases (150 cancer and 150 benign or normal) from DMIST

	CAD system A		CAD system B	
Standalone performance	75% sensitivity at 0.79 FPs/image		73% sensitivity at 0.77 FPs/image	
Radiologists	$N = 14$		$N = 15$	
	Without CAD	With CAD	Without CAD	With CAD
Average AUC	0.71	0.72	0.71	0.72
Average sensitivity	49%	51%	51%	53%
Average specificity	89%	87%	87%	86%

None of the changes were statistically significant. AUC = area under the receiver operating characteristic curve. N = number of radiologists in the study

readers were experienced breast radiologists and had been using CAD in their clinical practice. As summarized in Table 2, they found that the changes in the radiologists' sensitivity or specificity with CAD were only 1% to 2%. The standalone sensitivity of both CAD systems were 25% higher than the radiologists with or without CAD but had an average of more than 2 false positive marks per case. These results were very different from those observed in the early days of CAD development when radiologists were enthusiastic about CAD. They appeared to show that after radiologists used CAD in the clinic for a period of time, the many false positive CAD marks they have seen may have desensitized their attention and most of the marks were dismissed including true positives. In a screening setting, the time a radiologist has to spend to exclude over 2000 false positive marks in order to gain one or two cancers per 1000 examinations is considered not cost-effective by many radiologists. This study indicated that the specificity of a decision support tool has to be high to avoid inducing fatigue on clinicians' response to the computer's recommendations.

Although these clinical experiences of CAD were observed from screening mammography, they reveal the many challenges of implementing CAD or AI tools in the clinic and may provide some guidance on the development of the new generations of CAD for various applications in general. Accuracy and workflow efficiency are important considerations in clinical practice. User training is crucial to ensure their understanding of the limitations and capability of CAD and thus avoid improper use or disillusion. Clinicians' experiences and level of enthusiasm with CAD also strongly impact on whether they will accept a CAD tool and how they may respond to its recommendation. Performance standards and acceptance testing should be established to ensure the CAD tool can meet certain criteria before routine clinical use. Quality assurance is needed to monitor the consistency and accuracy of the CAD tool over time, as well as to prevent improper use that may impact patient safety. Fully automated medical decision systems are ideal, but experienced clinicians' supervision is vital as many clinical cases may not evolve following a statistical model and require human intelligence to determine the best course of action based on the individual patient's conditions and medical history. The AI community has recently scaled back the expectation and defines a less ambitious term as "narrow AI" or "Augmented Intelligence," recognizing the supporting role of machine learning algorithms. Regardless of CAD, narrow AI, or general AI, there are many challenges of developing machine-learning-based tools for medicine. Some of the challenges are discussed below.

Data Collection

The majority of the studies to date on the application of deep learning to medical imaging reported very promising results that often exceeded clinicians' performance, raising high expectations of AI tools. However, most of the studies used small training set and the trained models have not been subjected to rigorous validation with large real world test data. The generalizability of these deep learning models to new patients or to different clinical settings is still unknown.

One of the basic requirements to develop a robust machine learning algorithm is a sufficiently large training sample set with verified reference truth that are representative of the characteristics of the population of interest. Training deep learning is even more demanding because of the extremely large number of weights in a DCNN structure. Even with effective regularization methods to reduce overfitting, how general the feature representations it has learned still depends on how much the training set covers. AlexNet has over 60 million weights and the "ImageNet" data set for training includes over 1.2 million images with annotations. Sun et al. [31] showed that the performance of a DCNN increased linearly with the orders of magnitude of the training data and the performance of a DCNN with large learning capacity continued to increase even when the training set increased to over 300 million images.

Collection of medical imaging data that are representative of the patient population and with reliable annotation or reference truth is costly. While it is relatively easy to collect a large number of normal cases for a screening modality, it is difficult to collect sufficient abnormal cases, especially that the different classes in the data set ideally should be balanced. For example, for disease such as breast cancer that is the most prevalent cancer in women, there are only several cases per thousand in the screening population. It is difficult to collect enough breast cancer mammograms or tomosynthesis that can cover the variabilities in image features due to factors such as patient age, breast density and size, habitus, race, ethnicity, imaging protocols, and processing methods. The collection of normal and abnormal cases with special imaging modalities such as MR or PET is even more challenging because a relatively small number of patients will have these examinations and the availability may depend on the protocols for different types of diseases in different health systems.

Studies have demonstrated the feasibility of collecting a large number of annotated cases by data mining and natural language processing of the electronic medical record (EMR) [44] and clinical annotations in picture archiving and communication system (PACS) [45]. The accuracy and usefulness of the labels or annotations obtained from these methods not only depend on the methods used but also how the information is generated and stored in the systems. It has been shown that automatically mined disease labels or annotations can include substantial noise in a data set, as in the large public set of chest radiographs [46]. In the Digital Mammography DREAM Challenge (2016–2017) that aimed at building a model to help reduce the recall rate for breast cancer screening [47], the participants were provided with over 640,000 training mammograms from over 86,000 women. The training set only included breast-level labeling without lesion annotation. The winning teams all used deep learning approach but the highest performance only reached an area under the receiver operating characteristic curve (AUC) of 0.8744, and a sensitivity of 80% at specificity of 80.8%. The false positive rate, and thus potentially the recall rate, was much higher than that of an experienced breast radiologist at comparable sensitivity even for the top deep learning model. This example illustrates that, although the total number of images appeared to be large, the lack of high quality labeling may reduce its effectiveness in deep learning training. In general, weakly supervised training, unsupervised training, or using training set with substantial labeling errors is not as effective as supervised training with well-curated cases for the same training sample size; a much larger sample size is required to achieve similar performance for a DCNN model as a well-curated training set.

Data mining of the unstructured text and non-standardized reporting in current EMR or PACS systems is challenging, especially for more complex CAD task such as treatment response monitoring, in which a case may include multiple stages of diagnosis and treatment involving multiple imaging examinations and clinical tests. To generate reference standards for CAD development, one needs to correlate the imaging and clinical test data with outcomes at the various stages. It is a difficult process even if performed manually. Automation will be useful but it may require the development of an intelligent data mining tool. For patient cases that have been transferred between different hospitals, the incomplete prior or follow-up information may introduce errors into data curation. To facilitate collecting big data for development of AI towards precision medicine in the future, it will be prudent for the vendors and users to establish standardized reporting methods and structures among the various data archiving systems. In addition, establishing standardized protocols for secure electronic transmission of patient files among hospitals for referral patients will not only improve the health care of referral patients by transferring patient data accurately and efficiently, but also improve the accuracy of data mining for these cases. Ultimately, multi-institutional collaboration may be the best approach to building big database, which can cover the wide ranges of heterogeneous imaging protocols and equipment, clinical settings, and patient characteristics, to accelerate the development of robust deep learning models for each type of diseases that may be more readily applicable to various clinical environments.

Transfer Learning

Transfer learning is a common approach that deep learning developers use when the training set was small. In transfer learning, a DCNN that has been well trained with a large training set from a source domain is adapted to a new target task by fine-tuning the DCNN using a relatively small training set from the target domain. DCNN is considered a feature extractor that learns representation of the input data by extracting multiple levels of abstractions by its convolutional layers. Yosinski et al. [48] showed that the learned features in the shallow layers are more generic, whereas the learned features in the deeper layers become increasingly specific to the task that the DCNN is being trained for. Since the features are decomposed into numerous components in a DCNN, and most images are composed of some common basic elements, the knowledge learned by a trained DCNN in extracting features is shown to be transferrable to images from different domains. The transferability of features decreases as the differences between the source domain and the target domain increase. However, even for very different source and target tasks, transfer learning by initializing a DCNN with weights trained for another source task can outperform the same DCNN trained with randomly initialized weights for the target task.

For training deep learning models in medical imaging, the majority of studies used transfer learning due to the limited data available. To date, the largest annotated public data set available is the ImageNet data, which contained photographic images containing over 1000 classes of everyday life objects such as animals, vehicles, plants, ships, planes, etc. Most of the DCNN models in medical imaging were trained by transfer learning using models initialized with ImageNet-pretrained weights and fine-tuned by limited medical image data. Transfer learning was generally found to be useful in improving the training convergence and robustness of the DCNNs. In some cases, the pretrained DCNNs were used as feature extractor without fine-tuning; the deep features extracted from deploying the pretrained DCNN to the image data of the target domain were used as predictor variables to train an external classifier for the target task.

Although transfer learning can alleviate the problem of limited data to a certain degree, a large training set is still needed to achieve a high performance DCNN model for a given target task. Samala et al. [49] conducted a study to evaluate the effect of training set size on the performance of a transfer-trained DCNN for the target task of classifying malignant and benign breast masses in digital breast tomosynthesis (DBT). The ImageNet-pretrained AlexNet with 5 convolutional layers and 3 fully connected layers was appended with 2 additional fully connected layers (total of 5 fully connected layers) to reduce the classes from over 1000 to 2 (malignant and benign) and transfer-trained for the target task. Because the DBT data set was small and mammogram data were relatively abundant, the

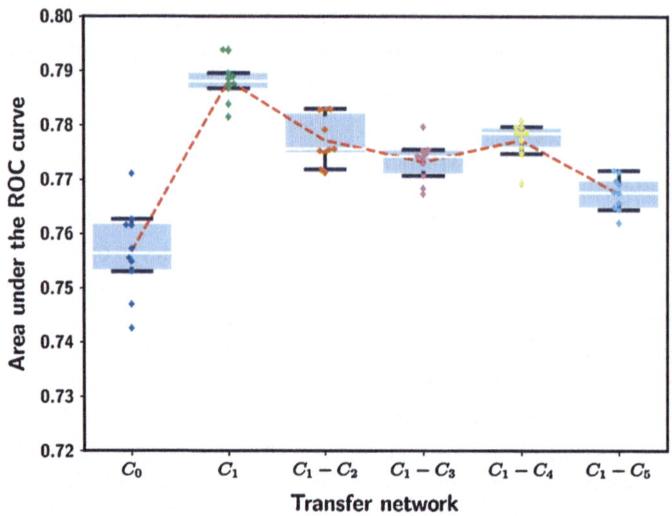

Fig. 2 The effect of different number of layers of the DCNN being frozen during transfer learning of ImageNet-pretrained AlexNet to classify malignant and benign masses on mammograms. The area under the receiver operating characteristic curve (AUC) for the test ROIs was plot as box-and-whisker plots of 10 repeated experiments under each condition. The training set and the test set consists of 12,360 and 7272 ROIs after augmentation, respectively. C_0 denotes no layer was frozen, i.e., the pretrained weights in all layers were allowed to be updated. C_1 denotes the first convolutional layer was frozen, C_1–C_i ($i = 2, 3, 4, 5$) denotes the C_1 to C_i convolutional layers were frozen during transfer training. The result shows that C_1-frozen training provided the best test AUC for this task (reprint with permission [49])

pretrained AlexNet was transfer-trained in the first stage for the classification of masses on mammograms, which brought the AlexNet from an unrelated classification task on non-medical ImageNet data to a task (mammography) much closer to the target task (DBT). A small DBT set was then used for a second-stage transfer training to the target task. Their mammography set contained 2242 unique views (craniocaudal or mediolateral oblique) with 2454 regions of interest (ROIs) containing breast masses. The DBT set contained 324 unique views with 1585 ROIs (5 slices or ROIs from each mass), which was partitioned by case into a training set of 1140 ROIs and an independent test set of 445 ROIs. Each ROI was flipped and rotated to obtain 8 augmented versions to reduce noise. To evaluate the training sample size effects on stage 1 and stage 2 fine-tuning, several transfer learning strategies were compared: (A) single-stage transfer learning with mammography data, in which the first convolutional layer (C_1) of AlexNet was frozen and all other layers were allowed to be fine-tuned, (B) two-stage transfer learning with mammography data in stage 1 and DBT training set in stage 2, in which C_1 was frozen in both stages, (C) two-stage transfer learning similar to (B) except that convolutional layers C_1 to F_4 were frozen in stage 2, and (D) single-stage transfer learning with DBT training set, in which C_1 was frozen, was also trained as a baseline for comparison. The results are summarized in Figs. 2, 3, 4, and 5.

Figure 2 shows the dependence of the test performance, in terms of AUC, of the transfer-trained AlexNet on the number of layers being frozen during transfer training for the classification of masses on mammograms. The AUC was the highest when only C_1 was frozen. However, if all layers were allowed to be re-trained (C_0), the transfer trained AlexNet did not perform well, probably because the mammography data was not large enough to fine-tune the large number of weights. Figure 3 shows the dependence of the test AUC on the sample size of the training mammography data. The test AUC was obtained by applying the AlexNet transfer-trained with mammography data directly to classify the masses on DBT with-

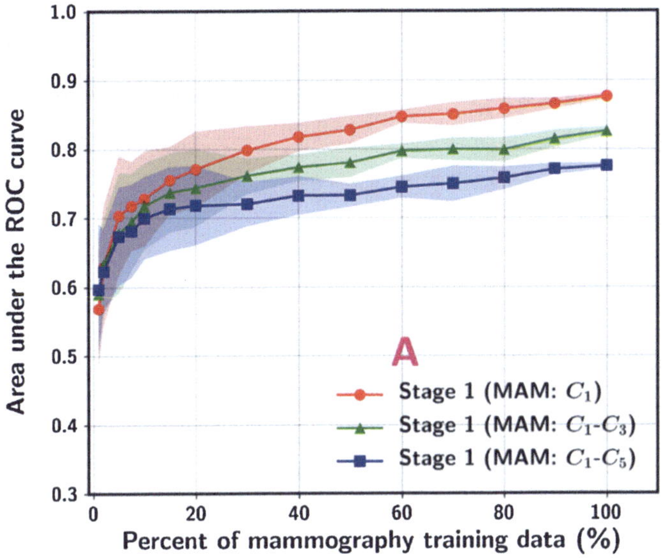

Fig. 3 Dependence of test AUC on mammography training sample size using strategy (A) transfer training. The varied training sample size was simulated by random drawing by case of a percentage (ranging from 1% to 100%) from the entire set of 19,632 mammography ROIs. The ROI-based AUC performance for classifying the 9120 DBT training ROIs (serve as a test set at this stage) for three transfer networks at Stage 1. The data point and the upper and lower range show the mean and standard deviation of the test AUC resulting from ten random samplings of the training set of a given size from the original set (reprint with permission [49])

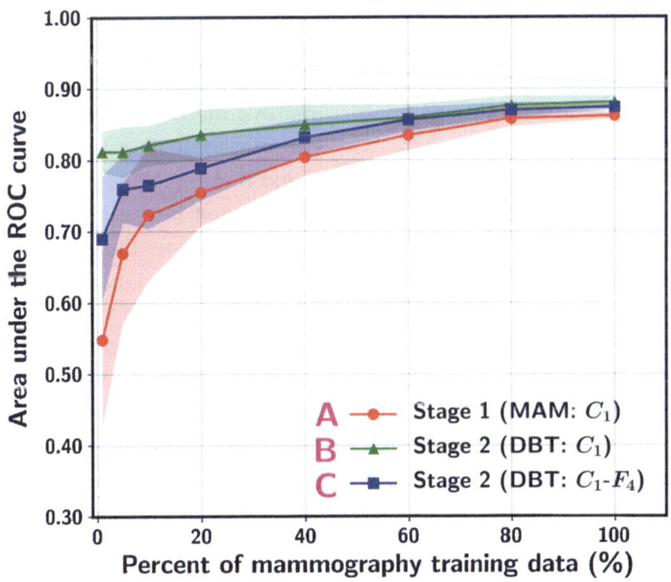

Fig. 4 ROI-based AUC on the DBT test set while varying the mammography sample size available for transfer training. The data point and the upper and lower range show the mean and standard deviation of the test AUC resulting from ten random samplings of the training set of a given size from the original set. "A. Stage 1 (MAM:C_1)" denotes single-stage training using mammography data and the C_1-layer frozen during transfer learning without stage 2. "B. Stage 2 (DBT:C_1)" denotes stage 2 C_1-frozen transfer learning at a fixed (100%) DBT training set size after Stage 1 transfer learning (curve A). "C. Stage 2 (DBT:C_1-F_4)" denotes Stage 2 C_1-to-F_4-frozen transfer learning at a fixed (100%) DBT training set size after stage 1 transfer learning (curve A) (reprint with permission [49])

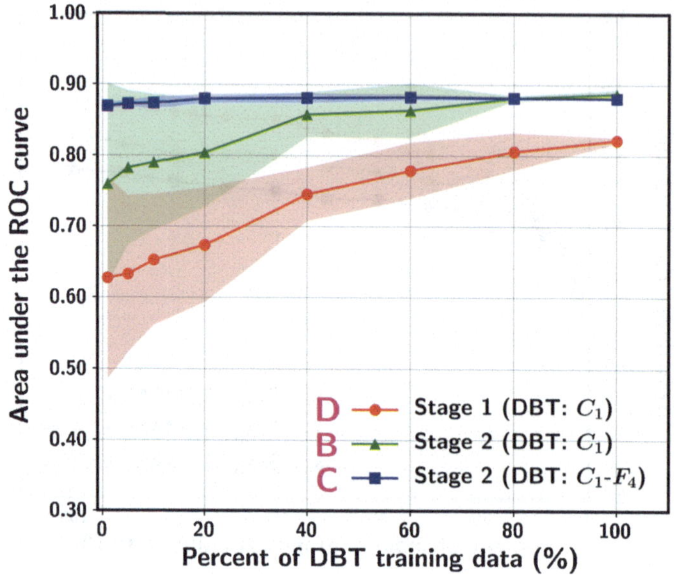

Fig. 5 ROI-based AUC on the DBT test set while varying the simulated DBT sample size available for transfer training. The data point and the upper and lower range show the mean and standard deviation of the test AUC resulting from ten random samplings of the training set of a given size from the original set. "D. Stage 1 (DBT:C_1)" denotes single-stage training using DBT data with the C_1-layer frozen during transfer learning without Stage 2. "B. Stage 2 (DBT:C_1)" denotes Stage 2 C_1-frozen transfer learning after Stage 1 transfer learning with a fixed (100%) mammography training set. "C. Stage 2 (DBT:C_1-F_4)" denotes Stage 2 C_1-to-F_4-frozen transfer learning after Stage 1 transfer learning with a fixed (100%) mammography training set (reprint with permission [49])

out the second-stage fine-tuning with DBT. The test AUC increased steadily as the training sample size increased. For a given training set size, the test AUC decreased as more and more layers were frozen, indicating that the learning capacity of the DCNN was restricted and insufficient knowledge was learned from the mammography data. Furthermore, the test AUC on the DBT set (Fig. 3) was higher than that on the mammography test set (Fig. 2) at the corresponding training sample size and frozen layers, indicating that mammography is an effective auxiliary domain for transfer training to DBT and that malignant and benign masses in DBT are easier to be distinguished by DCNN, similar to that by human vision.

Figures 4 and 5 compared the two-stage transfer learning to one-stage transfer learning on the classification of masses in the DBT test set. Several observations can be made. First, the test AUC increases with training sample size either in stage 1 or stage 2. Second, when the training set in the target domain is small, the additional stage of pre-training with data of auxiliary domain can improve the overall performance at all training sample sizes in the range studied (compare curves A and B in Fig. 4, and curves B and D in Fig. 5). Third, when too many layers are frozen during transfer learning, the performance of the DCNN after two-stage training may not reach the same level as that of the DCNN with less layers frozen using the same training sample sizes (compare curves B and C in Fig. 4), indicating that the DCNN cannot learn adequately from the training data if the learning capacity of the DCNN is overly restricted. Fourth, on the other hand, when the DCNN is well trained in the source domain and the training set in the target domain is very small, freezing most of the layers during transfer training may be beneficial by avoiding the loss of the pretrained knowledge without adequate learning from the target domain data (compare the small sample size region of curves B and C in Fig. 5). Although one can expect that the training sample size required for transfer training for a given task will depend on many factors such as the complexity of the tasks and the DCNN structure, the differences in the characteristics between the source and the target domains, the relative training sample sizes between the tasks, the relative trends observed from this study will likely be applicable to many transfer learning applications, and multi-stage transfer training with data from similar domains should be helpful if the training data of the target domain is too scarce.

Data Augmentation

Data augmentation generates multiple slightly different versions of images from each image in the original training set. Data augmentation may use techniques such as flipping the image in various directions, translating the image within a range of distance, cropping the image in different ways, rotating the image within a range of angles, scaling the image over a range of factors, generating shape- and intensity-transformed images by linear or non-linear methods. Data augmentation can be implemented on-line or off-line and an augmentation operation in a specified range can be performed randomly or by fixed increments. For off-line augmentation, the augmented versions of the images are pre-generated and mixed with the original data into a larger training set, which is randomly grouped as mini-batches for the DCNN training. If the various techniques are applied in combinations, the apparent number of training images can increase easily to hundreds or thousands of times. For on-line augmentation, the various augmentation techniques are usually implemented as a part of the DCNN pipeline with user-selectable probability and range. The original training set is input in mini-batches but each image in a batch is randomly altered according to the pre-selected probability and range of the augmentation techniques. The number of times an image is augmented in a given training run will depend on the number of training epochs chosen and the pre-selected probabilities for the different augmentation techniques. The choice between off-line and on-line augmentation may depend on the tradeoffs between computational resources and storage space or memory; off-line augmentation is more practical if the available training set is small as it requires more space and memory for the augmented set, while on-line augmentation is preferred for large training sets if computational resource is plentiful. Data augmentation introduces variations or jittering to the original data, thereby reducing the risk of overfitting to a small training set and improving generalizability [28, 50, 51]. However, it is important to note that augmenting the training set to a certain size is not equivalent to having a set of independent training samples of comparable size. Since the features in the augmented versions of an image are highly correlated and the CNN learning is invariant to many of these small variations, the augmented images do not provide much new knowledge for the DCNN to learn in comparison to new independent images. In particular, if the original small training set does not include representative samples of certain characteristics, data augmentation will not generate samples of the missing types for the DCNN training; for example, if there are no spiculated nodules in the original data, augmentation cannot generate spiculated nodules for the DCNN to learn. Other more sophisticated data augmentation methods are also being considered, such as generative adversarial networks (GANs) that can generate images with mixed features learned from different images after training on the available sample images [52], digitally generate artificial lesions inserted into normal images [53, 54], or inserting real lesions to other locations of normal or abnormal images [55]. Investigations are needed to evaluate issues such as how effective the data augmentation methods are compared to one another for a given sample size, whether the features learned from the artificial lesions, especially texture features, may help or hinder DCNN learning of real features, whether real lesions inserted at other locations can contribute new features to learn, and whether the usefulness of the augmented data for deep learning depends on the target task.

Training, Validation, and Independent Testing

During training of a machine learning model including deep learning, a validation set is generally used for guiding the optimization of the parameters. The validation set is used to compare the performances of models with different parameter sets, or to monitor the changes in the performance or cost during iterative training of the model weights. The validation set may be split from the training set by cross validation or by hold-out. Regardless of the methods, the validation set is a part of the training process because it is repeatedly used to guide training, and the model structure and parameters are usually chosen to maximize the performance on the validation set. It is well known in machine learning that the training or validation performance is generally optimistically biased [56–61]. To estimate the true performance of the trained model in unknown cases, one has to use an independent test set that has not been seen by the model in the training process and is representative of the population to which the trained model will be applied. To date, most of the published studies only include cross validation results, and even in studies with a "hold-out" test set, the test set will be turned into a validation set if the same test set is used for evaluation many times during model development and eventually the best model is chosen based on the performance of the test set. The American Association of Physicists in Medicine (AAPM) CAD Subcommittee (renamed as Computer-Aided Image Analysis Subcommittee in 2018) has published an opinion paper to discuss the training and evaluation methodology for development of CAD systems [62]. The importance and the strategy of collecting a representative independent test set and the potential biases on the reported "test" result due to multiple repeated use of the same test set are discussed in more details. Deep-learning-based CAD or AI follows similar general principles as conventional machine learning methods, and the need for independent testing will be even more important due to the vast capacity of deep learning to extract and memorize information from the training set.

Acceptance Testing, Preclinical Testing, and User Training

If properly trained with a large data set, deep learning is expected to be more robust and more accurate than conventional machine learning approaches. However, studies showed that deep learning in medical imaging, or machine learning in general, can learn non-medical features that are not related to the medical conditions of the patient but other properties such as image acquisition protocols or equipment, image processing techniques, or even other markings and accessories related to the facilities or patient comorbidity that are recorded in the images [63]. As a result, a deep learning algorithm well trained and independently tested showing high accuracy using data collected from the same site(s) may not be generalizable to different clinical sites that may have different population or imaging characteristics. Even if a CAD or AI algorithm is approved by FDA for clinical use, a clinical site should conduct acceptance testing, similar to the installation of a new medical device or equipment, using a set of representative local data to verify that its performance for the local patient population can pass a certain standard or reference level before clinical implementation. In addition, after the AI tool is implemented in the clinical workflow, the users should allow for a test period in which they refrain from being influenced by the CAD output. The users should familiarize themselves with the output of the CAD tool and quantitatively, if possible, assess the performance of the CAD tool on a large number of consecutive clinical cases. The users should evaluate critically the strengths and weaknesses of the CAD tool based on follow-up review of the outcomes of the cases, so as to recognize the characteristics of cases that the CAD tool makes mistakes or the CAD tool makes correct recommendations whereas the clinician may have failed. The hands-on experience of the performance of the CAD tool will allow the users to learn how to reduce the risk of accepting erroneous recommendation while taking advantage of the recommendations for cases that the CAD tool is useful. The test period will serve both as a real world evaluation of the CAD tool on the local population and user training. With better understanding of the AI's limitation and capability, the users may be able to establish proper expectation and confidence level on the CAD tool and thus reducing the risk of improper use or negative outcomes of using CAD.

Quality Assurance and Performance Monitoring

With the new generation of CAD, there are high expectations that they will be far more robust than the conventional CAD systems, especially that many of the studies reported performance higher than those of clinicians. Although the initial concerns of AI algorithms replacing radiologists have tamped down, the expectations of using AI to improve workflow efficiency or reduce workload are prevalent. A recent observer study [64] compared breast cancer detection in DBT by radiologist alone to radiologist using deep-learning-based CAD as a concurrent reader that marked suspected lesions and showed the confidence of malignancy on the DBT slices. A data set of 260 DBT cases including 65 cancer, 65 benign, and 130 normal cases were read by 24 radiologists. The experimental concurrent CAD had a case-based sensitivity of over 90% and a specificity of over 40%, which are higher than all of the CAD tools currently used in screening DM. They demonstrated that reading with CAD could provide all the benefits a radiologist would hope for: reducing the average reading time by more than 50% for a DBT case, increasing sensitivity and specificity, as well as reducing recall rate. In another study [65], researchers developed a DCNN to identify normal mammograms from screening cases. With 10-fold cross validation, they showed that the DCNN could identify 34% and 91% of the normal mammograms at a negative predictive value (NPV) of 0.99 for a cancer prevalence of 15% and 1%, respectively. The study showed the potential of using DCNN to improve radiologists' workflow efficiency by excluding the negative mammograms from reading.

For an AI model to be a useful routine clinical tool, it is crucial to validate that its performance in clinical settings can meet certain standards and is consistent over time, similar to other medical devices, especially for any AI model that is designed to operate as a decision maker, rather than as a decision support tool or a second opinion. The acceptance testing or preclinical testing described above can serve as the baseline performance on the local population. Since the performance of DCNN is affected by the properties of the input images, which may be determined by a number of factors such as the imaging techniques or equipment and the image processing or reconstruction software or parameters that may change intentionally or unintentionally due to many factors, periodic quality assurance (QA) procedures should be established to monitor the performance of the CAD tool as well as the performance of clinicians using CAD over time. The AAPM CAD Subcommittee has published an opinion paper on the quality assurance and user training on CAD devices in clinical use [66]. The discussions have not attracted much attention, probably because of the limited use of CAD in the clinic at that time. Currently FDA has no post-market monitoring and regulations on the consistency or accuracy of CAD software as second opinion in clinical use after it is approved and there is no control of off-label use. As CAD/AI tools are anticipated to have widespread use in health care in the future, either as second opinion or automated decision maker in some applications such as pre-screening or triaging, their impact on patient care or welfare can be much greater. It will be important for organizations such as the American College of Radiology (ACR), the Radiological Society of North America (RSNA), the European Society of Radiology, and AAPM to provide leadership to establish performance standards, QA and monitoring procedures, and compliance guidance, to ensure safety and effectiveness for implementation and operation of CAD tools in clinical practice.

Interpretability of CAD/AI Recommendations

The DCNN learns multiple levels of feature representations from the input data by using the deep architecture of convolutional layers. At present a DCNN model is mostly operated like a black-box as there is no easy way to explain how and what the DCNN has learned to perform a specific classification task. Researchers have developed methods to visualize the feature maps at each convolutional layer [67, 68] and to highlight the target objects recognized by the DCNN with a class activation map [69]. The feature maps illustrate the deep features [70] extracted by the DCNN and the class activation map may be correlated with the target location or the locations of the most important features for classification. These visualization tools are the first steps to explore the inner workings of deep learning but they are still far from being able to translate the deep learning output to interpretable clinical decisions, especially for tasks more complex than lesion detection. For CAD/AI to be more widely acceptable as a clinical decision support tool, it should be able to more intelligently present the recommendation to clinicians with reasons, correlating the findings with the medical conditions and data of the patient, and ideally, be able to present further explanations if the clinician has questions on the recommendation. Uncovering the relationship between the machine findings with medical conditions of the patient or even utilizing deep learning and big data analytics to discover new links between disease and clinical data or symptoms will be an important area of research to enable CAD to deliver interpretable diagnosis to clinicians and advance CAD towards true AI in medicine.

Summary

Deep learning is expected to revolutionize CAD and image analysis in medicine. Although machine learning has been applied to CAD and medical image analysis for over three decades, CAD has not been commonly used in the clinic due to the limited performance of conventional machine learning approaches. The recent success of deep learning technology spurs new efforts to develop CAD or AI tools for many applications in health care. Numerous studies have reported promising results. Amid the high expectations of the accuracy and efficiency that AI can bring to medicine, many challenges have yet to be overcome in order to integrate the new generation of CAD tools into clinical practice and to minimize the risk of unintended harm to patients. The discussion in this chapter is not limited to computer-aided lesion detection. Similar considerations are applicable to any CAD tools in general, such as those for disease characterization, staging, treatment planning, surgical guidance, treatment response assessment, recurrence monitoring, and prognosis or survival prediction. Big databases have to be collected to provide sufficient training and validation samples to develop robust deep learning models and independent testing with internal and external multi-institutional data to assess generalizability; performance standards, acceptance testing, and quality assurance procedures should be established for each type of applications to ensure the performance of a deep learning model can meet the requirements in the local clinical environment and remains consistent over time; adequate user training in local patient population is vital to allow users to understand the capability and limitations of the CAD tool, establish realistic expectations, and avoid improper use or disillusion; CAD recommendation has to be interpretable to allow clinicians to make informed decisions. More importantly, workflow efficiency and costs are major considerations in health care. A decision support tool will not be acceptable if it requires additional time and/or costs without significant clinical benefits. It is important for CAD researchers and developers to understand the preferred mode of assistance by clinicians for each type of clinical tasks, design effective CAD tools, and deliver interpretable outputs by taking into consideration the practical issues in clinical settings. If properly developed, validated, and implemented, it can be expected that the efficient data analytics from CAD or AI tools can complement the human intelligence of clinicians to improve the accuracy and workflow and thus patient care.

Acknowledgment This work is supported by National Institutes of Health award number R01 CA214981.

Disclosures The authors have no conflicts to disclose.

References

1. Winsberg F, Elkin M, Macy J, Bordaz V, Weymouth W (1967) Detection of radiographic abnormalities in mammograms by means of optical scanning and computer analysis. Radiology 89:211–215
2. Kimme C, O'Laughlin BJ, Sklansky J (1977) Automatic detection of suspicious abnormalities in breast radiographs. Data structures, computer graphics and pattern recognition. Academic Press, New York
3. Spiesberger W (1979) Mammogram inspection by computer. IEEE Trans Biomed Eng 26:213–219
4. Semmlow JL, Shadagopappan A, Ackerman LV, Hand W, Alcorn FS (1980) A fully automated system for screening mammograms. Comput Biomed Res 13:350–362
5. Doi K (2015) Chapter 1. Historical overview. In: Li Q, Nishikawa RM (eds) Computer-aided detection and diagnosis in medical imaging. Taylor & Francis Group, LLC, CRC Press, Boca Raton, FL, pp 1–17
6. Chan H-P, Doi K, Galhotra S, Vyborny CJ, MacMahon H, Jokich PM (1987) Image feature analysis and computer-aided diagnosis in digital radiography. 1. Automated detection of microcalcifications in mammography. Med Phys 14:538–548
7. Chan H-P, Doi K, Vyborny CJ, Schmidt RA, Metz CE, Lam KL et al (1990) Improvement in radiologists' detection of clustered microcalcifications on mammograms. The potential of computer-aided diagnosis. Investig Radiol 25:1102–1110
8. LeCun Y, Bengio Y, Hinton G (2015) Deep learning. Nature 521:436–444
9. Fukushima K, Miyake S (1982) Neocognitron: a new algorithm for pattern recognition tolerant of deformations and shifts in position. Pattern Recogn 15:455–469
10. LeCun Y, Boser B, Denker JS, Henderson D, Howard RE, Hubbard W et al (1990) Handwritten digit recognition with a back-propagation network. Proc Adv Neural Inf Process Syst:396–404
11. Lo SCB, Lin JS, Freedman MT, Mun SK (1993) Computer-assisted diagnosis of lung nodule detection using artificial convolution neural network. Proc SPIE 1898:859–869
12. Lo SCB, Chan H-P, Lin JS, Li H, Freedman M, Mun SK (1995) Artificial convolution neural network for medical image pattern recognition. Neural Netw 8:1201–1214
13. Chan H-P, Lo SCB, Helvie MA, Goodsitt MM, Cheng SNC, Adler DD (1993) Recognition of mammographic microcalcifications with artificial neural network. Radiology 189(P):318
14. Chan H-P, Lo SCB, Sahiner B, Lam KL, Helvie MA (1995) Computer-aided detection of mammographic microcalcifications: pattern recognition with an artificial neural network. Med Phys 22:1555–1567
15. Chan H-P, Sahiner B, Lo SCB, Helvie MA, Petrick N, Adler DD et al (1994) Computer-aided diagnosis in mammography: detection of masses by artificial neural network. Med Phys 21:875–876
16. Sahiner B, Chan H-P, Petrick N, Wei D, Helvie MA, Adler DD et al (1995) Image classification using artificial neural networks. Proc SPIE 2434: 838–845
17. Wei D, Sahiner B, Chan H-P, Petrick N (1995) Detection of masses on mammograms using a convolution neural network. In: Proceedings of the International Conference on Acoustics, Speech and Signal Processing. 95CH2431-5, pp 3483–3486
18. Sahiner B, Chan H-P, Petrick N, Wei D, Helvie MA, Adler DD et al (1996) Classification of mass and normal breast tissue: a convolution neural network classifier with spatial domain and texture images. IEEE Trans Med Imaging 15:598–610
19. Zhang W, Doi K, Giger ML, Wu Y, Nishikawa RM, Schmidt RA (1994) Computerized detection of clustered microcalcifications in digital mammograms using a shift-invariant artificial neural network. Med Phys 21:517–524
20. Hinton GE, Osindero S, Teh Y-W (2006) A fast learning algorithm for deep belief nets. Neural Comput 18(7):1527–1554
21. Bengio Y, Lamblin P, Popovici D, Larochelle H (2006) Greedy layer-wise training of deep networks. Proc Adv Neural Inf Process Syst 19:153–160
22. Erhan D, Bengio Y, Courville A, Manzagol P-A, Vincent P, Bengio S (2010) Why does unsupervised pre-training help deep learning? J Mach Learn Res 11:625–660
23. Nair V, Hinton GE (2010) Rectified linear units improve restricted boltzmann machines. In: Proceedings of the 27th International Conference on Machine Learning, pp 807–814
24. Glorot X, Bordes A, Yoshua B (2011) Deep sparse rectifier neural networks. In: Proceedings of the 14th International Conference on Artificial Intelligence and Statistics, pp 315–323

25. Ranzato MA, Huang FJ, Boureau Y-L, LeCun Y (eds) (2007) Unsupervised learning of invariant feature hierarchies with applications to object recognition. In: IEEE Conference on Computer Vision and Pattern Recognition (CVPR), 17–22 June 2007. Minneapolis, MN, USA
26. Srivastava N, Hinton G, Krizhevsky A, Sutskever I, Salakhutdinov R (2014) Dropout: a simple way to prevent neural networks from overfitting. J Mach Learn Res 15:1929–1958
27. Ioffe S, Szegedy C (2015) Batch normalization: accelerating deep network training by reducing internal covariate shift. In: Proceedings of the 32nd International Conference on Machine Learning (ICML'15), vol 37, pp 448–456. arXiv:1502.03167
28. Krizhevsky A, Sutskever I, Hinton GE (2012) ImageNet classification with deep convolutional neural networks. Adv Neural Inf Process Syst:1097–1105
29. Russakovsky O, Deng J, Su H, Krause J, Satheesh S, Ma S et al (2015) Imagenet large scale visual recognition challenge. Int J Comput Vis 115(3): 211–252
30. He K, Zhang X, Ren S, Sun J (2015) Deep residual learning for image recognition. In: Proceedings of the IEEE Conference on Computer Vision and Pattern Recognition (CVPR), pp 770–778. arXiv:1512.03385
31. Sun C, Shrivastava A, Singh S, Gupta A (2017) Revisiting unreasonable effectiveness of data in deep learning era. arXiv:1707.02968
32. Litjens G, Kooi T, Bejnordi BE, Setio AAA, Ciompi F, Ghafoorian M et al (2017) A survey on deep learning in medical image analysis. Med Image Anal 42:60–88
33. Sahiner B, Pezeshk A, Hadjiiski LM, Wang X, Drukker K, Cha KH et al (2019) Deep learning in medical imaging and radiation therapy. Med Phys 46(1):e1–e36
34. Mazurowski MA, Buda M, Saha A, Bashir MR (2019) Deep learning in radiology: an overview of the concepts and a survey of the state of the art with focus on MRI. J Magn Reson Imaging 49(4):939–954
35. Fauw JD, Ledsam JR, Romera-Paredes B, Nikolov S, Tomasev N, Blackwell S et al (2018) Clinically applicable deep learning for diagnosis and referral in retinal disease. Nat Med 24(9):1342–1350
36. Janowczyk A, Madahushi A (2016) Deep learning for digital pathology image analysis: a comprehensive tutorial with selected use cases. J Pathol Inform 7:29
37. Chan H-P, Hadjiiski LM, Samala RK (2019) Computer-aided diagnosis in the era of deep learning. Med Phys (accepted)
38. Cole EB, Zhang Z, Marques HS, Hendrick RE, Yaffe MJ, Pisano ED (2014) Impact of computer-aided detection systems on radiologist accuracy with digital mammography. AJR Am J Roentgenol 203:909–916
39. Gilbert FJ, Astley SM, Gillan MGC, Agbaje OF, Wallis MG, James J et al (2008) Single reading with computer-aided detection for screening mammography. N Engl J Med 359(16):1675–1684
40. Gromet M (2008) Comparison of computer-aided detection to double reading of screening mammograms: review of 231,221 mammograms. Am J Roentgenol 190(4):854–859. https://doi.org/10.2214/ajr.07.2812
41. Taylor P, Potts HWW (2008) Computer aids and human second reading as interventions in screening mammography: two systematic reviews to compare effects on cancer detection and recall rate. Eur J Cancer 44:798–807
42. Fenton JJ, Abraham L, Taplin SH, Geller BM, Carney PA, D'Orsi C et al (2011) Effectiveness of computer-aided detection in community mammography practice. J Natl Cancer Inst 103(15):1152–1161. https://doi.org/10.1093/jnci/djr206
43. Lehman CD, Wellman RD, Buist DSM, Kerlikowske K, Tosteson ANA, Miglioretti DL (2015) Diagnostic accuracy of digital screening mammography with and without computer-aided detection. JAMA Intern Med 175(11):1828–1837
44. Zech J, Pain M, Titano J, Badgeley M, Schefflein J, Su A et al (2018) Natural language-based machine learning models for the annotation of clinical radiology reports. Radiology 287(2):570–580. https://doi.org/10.1148/radiol.2018171093
45. Yan K, Wang X, Lu L, Summers RM (2018) DeepLesion: automated mining of large-scale lesion annotations and universal lesion detection with deep learning. J Med Imaging 5(3):036501
46. Oakden-Rayner L (2017) Exploring the ChestXray14 dataset: problems.https://lukeoakdenrayner.wordpress.com/2017/12/18/the-chestxray14-dataset-problems/
47. The Digital mammograph DREAM Challenge (2017). https://www.synapse.org/#!Synapse:syn4224222/wiki/434547
48. Yosinski J, Clune J, Bengio Y, Lipson H (2014) How transferable are features in deep neural networks? In: Proceedings of the Advances in neural information processing systems (NIPS'14), pp 3320–3328
49. Samala RK, Chan H-P, Hadjiiski LM, Helvie MA, Richter CD, Cha K (2019) Breast cancer diagnosis in digital breast tomosynthesis: effects of training sample size on multi-stage transfer learning using deep neural nets. IEEE Trans Med Imaging 38(3): 686–696
50. Taylor L, Nitschke G (2017) Improving deep learning using generic data augmentation. arXiv:1708.06020
51. Wang J, Perez L (2017) The effectiveness of data augmentation in image classification using deep learning. arXiv:1712.04621
52. Goodfellow IJ, Pouget-Abadie J, Mirza M, Xu B, Warde-Farley D, Ozairy S et al (2014) Generative Adversarial Nets. arXiv:1406.2661v1
53. Badano A, Graff CG, Badal A et al (2018) Evaluation of digital breast tomosynthesis as replacement of full-field digital mammography using an in silico imaging trial. JAMA Network Open 1(7):e185474. https://doi.org/10.1001/jamanetworkopen.2018.5474

54. Cha KH, Petrick N, Pezeshk A, Graff CG, Sharma D, Badal A et al (2019) Reducing overfitting of a deep learning breast mass detection algorithm in mammography using synthetic images. Proc SPIE 10950:1095004
55. Pezeshk A, Petrick N, Chen W, Sahiner B (2017) Seamless lesion insertion for data augmentation in CAD training. IEEE Trans Med Imaging 36(4):1005–1015
56. Efron B (1983) Estimating the error rate of a prediction rule: improvement on cross-validation. J Am Stat Assoc 78:316–331
57. Fukunaga K, Hayes RR (1989) Effects of sample size on classifier design. IEEE Trans Pattern Anal Mach Intell 11:873–885
58. Hastie T, Tibshirani R, Friedman J (2001) The elements of statistical learning. Springer-Verlag, New York
59. Bishop CM (1995) Neural networks for pattern recognition. Clarendon Press, Oxford
60. Chan H-P, Sahiner B, Wagner RF, Petrick N (1999) Classifier design for computer-aided diagnosis: effects of finite sample size on the mean performance of classical and neural network classifiers. Med Phys 26:2654–2668
61. Sahiner B, Chan H-P, Petrick N, Wagner RF, Hadjiiski LM (2000) Feature selection and classifier performance in computer-aided diagnosis: the effect of finite sample size. Med Phys 27:1509–1522
62. Petrick N, Sahiner B, Armato SG, Bert A, Correale L, Delsanto S et al (2013) Evaluation of computer-aided detection and diagnosis systems. Med Phys 40(8):087001. https://doi.org/10.1118/1.4816310
63. Zech JR, Badgeley MA, Liu M, Costa AB, Titano JJ, Oermann EK (2018) Variable generalization performance of a deep learning model to detect pneumonia in chest radiographs: a cross-sectional study. PLOS Med 15(11):e1002683
64. Conant EF, Toledano AY, Periaswamy S, Fotin SV, Go J, Hoffmeister JW et al (2018) Improving accuracy and efficiency with concurrent use of artificial intelligence for digital breast tomosynthesis screening. In: Radiological Society of North America Scientific Assembly and Annual Meeting. RC215-14
65. Kyono T, Gilbert FJ, van der Schaar M (2019) Improving workflow efficiency for mammography using machine learning. J Am Coll Radiol. https://doi.org/10.1016/j.jacr.2019.05.012
66. Huo ZM, Summers RM, Paquerault S, Lo J, Hoffmeister J, Armato SG et al (2013) Quality assurance and training procedures for computer-aided detection and diagnosis systems in clinical use. Med Phys 40(7):077001. https://doi.org/10.1118/1.4807642
67. Zeiler MD, Fergus R (2014) Visualizing and understanding convolutional networks. In: Lecture Notes in Computer Science Computer Vision – European Conference on Computer Vision (ECCV) 2014, vol 8689, pp 818–833
68. Yosinski J, Clune J, Nguyen A, Fuchs T, Lipson H (2015) Understanding neural networks through deep visualization. arXiv:1506.06579v1
69. Zhou B, Khosla A, Lapedriza A, Oliva A, Torralba A (2016) Learning deep features for discriminative localization. In: Proceedings of the IEEE Conference on Computer Vision and Pattern Recognition (CVPR), pp 2921–2929
70. Samala RK, Chan H-P, Hadjiiski LM, Helvie MA, Cha KH, Richter CD (2017) Multi-task transfer learning deep convolutional neural network: application to computer-aided diagnosis of breast cancer on mammograms. Phys Med Biol 62:8894–8908

Medical Image Synthesis via Deep Learning

Biting Yu, Yan Wang, Lei Wang, Dinggang Shen, and Luping Zhou

Abstract

Medical images have been widely used in clinics, providing visual representations of under-skin tissues in human body. By applying different imaging protocols, diverse modalities of medical images with unique characteristics of visualization can be produced. Considering the cost of scanning high-quality single modality images or homogeneous multiple modalities of images, medical image synthesis methods have been extensively explored for clinical applications. Among them, deep learning approaches, especially convolutional neural networks (CNNs) and generative adversarial networks (GANs), have rapidly become dominating for medical image synthesis in recent years. In this chapter, based on a general review of the medical image synthesis methods, we will focus on introducing typical CNNs and GANs models for medical image synthesis. Especially, we will elaborate our recent work about low-dose to high-dose PET image synthesis, and cross-modality MR image synthesis, using these models.

Keywords

Deep learning · Medical image synthesis · Machine learning · Convolutional neural networks (CNNs) · Generative adversarial networks (GANs) · Magnetic resonance imaging (MRI) · Positron emission tomography (PET) · Brain

B. Yu · L. Wang
School of Computing and Information Technology, University of Wollongong, Wollongong, NSW, Australia

Y. Wang
School of Computer Science, Sichuan University, Chengdu, China

D. Shen
IDEA Lab, Department of Radiology and BRIC, University of North Carolina at Chapel Hill, Chapel Hill, NC, USA

L. Zhou (✉)
School of Electrical and Information Engineering, University of Sydney, Camperdown, NSW, Australia
e-mail: luping.zhou@sydney.edu.au

Introduction

As a technology to produce the visual representations of anatomical and pathological structures and their functions in human body, medical imaging is widely applied in clinics for disease diagnosis and treatment planning. It consists of various imaging protocols which have their specific insights to produce different modality images. For example, computed tomography (CT)

G. Lee, H. Fujita (eds.), *Deep Learning in Medical Image Analysis*, Advances in Experimental Medicine and Biology 1213,
https://doi.org/10.1007/978-3-030-33128-3_2

creates the images of internal organs, bones, soft tissue, and blood vessels to show electron density and physical density [22]; magnetic resonance imaging (MRI) provides diverse contrasts of soft tissues through setting different scanning parameters [33]; positron emission tomography (PET) enables the visualization of metabolic processes of scanned body part [12]. Due to their different and sometimes complementary characteristics, multiple modalities are usually utilized in the analysis of clinical applications. However, the acquisition of some modality images, such as PET and CT, increases the risks of radiation exposure, especially when patients have to take these imaging scanning multiple times during the entire treatment [37]. Thus, the high-quality and human-safe medical images are not easy to acquire in the practical applications. Besides, due to the different imaging protocols and the cost of multi-modality image acquisition, sufficient and consistent modalities are not always accessible for every patient [57]. To handle these issues, medical image synthesis, which is defined as an approach to modeling a mapping from the given source images to the unknown target images, has been widely explored by researchers [18].

Medical image synthesis has been used in various applications, e.g., estimation of missing images [43], knowledge transformation across modalities [53], image super-resolution [24], and annotated dataset creation [14]. Here, according to its applications, we roughly classify medical image synthesis into two main categories, i.e., within-modality synthesis and cross-modality synthesis. Specially, the within-modality synthesis usually aims at generating the higher-quality images from the input within-modality images of relatively lower quality. In contrast, the cross-modality synthesis targets to capture the useful structuring information in the source-modality to generate the target-modality image. Although these two categories are applied in different practical tasks, the underlying synthesis principles are similar. The conventional synthesis approaches exploit diverse nonlinear models, e.g., dictionary learning [45] and random forest [26], to process the handcrafted medical image features which are manually selected by professional experts during the synthesis. However, these handcrafted features have limited power to represent the complex visual information in medical images and therefore adversely affect the synthesis performance. Recently, deep learning based methods have mitigated this issue through automatically learning the task-specific features having sufficient descriptive power with the training of the mapping models [41, 55]. Through designing advanced deep learning models, the performance of medical image synthesis has been greatly improved.

In Table 1, a list of works that utilized deep learning models for medical image synthesis are presented. Here, we mainly focus on the synthesis applications for three major imaging modalities, i.e., CT, MR, and PET. The timeline for the development of these methods is summarized in Fig. 1. As shown in Table 1 and Fig. 1, deep learning approaches started to be popular for medical image synthesis in 2015 [42]. After two years of exploration, a large category of models, especially deep convolutional neural networks (CNNs) based architectures, became dominating for both within-modality and cross-modality synthesis in 2017 [10, 11, 20, 28, 31, 32, 51]. Before the end of 2017, a novel family of CNN based models, i.e., generative adversarial networks (GANs), attracted the attention of researchers and achieved promising results [3, 4, 7]. In 2018, more complicated CNN models were further explored in the conventional way [8,9,58]. At the same time, numerous GAN models with different frameworks were proposed in 2018 and 2019, and this research trend becomes more and more popular now.

In the rest of this chapter, we first discuss two typical types of deep learning models for medical image synthesis in section "Deep Learning Models for Medical Image Synthesis". Following that, we introduce four of our recent works for within-modality and cross-modality synthesis, respectively, in sections "Within-Modality Synthesis" and "Cross-Modality Synthesis." Finally, a brief conclusion about this chapter is given in section "Conclusion."

Table 1 Medical image synthesis publications with deep learning models

Publication	Method	Dataset	Organ
Within-modality synthesis			
CT (low-dose to full-dose)			
Chen et al. [11]	Custom three layer-CNN	NBIA[a]	Multiple body parts
Chen et al. [10]	CNN based residual encoder-decoder	NBIA[a]	Multiple body parts
Kang et al. [28]	CNN for wavelet domain denoising	low-dose CT[b]	Head, chest and abdomen
MRI (super-resolution or 3T to 7T)			
Zend et al. [58]	Residual CNN	Brainweb[c] NAMIC[d]	Brain
Chaudhari et al. [9]	Residual CNN	OAI [39]	Knee
Nie et al. [37]	Cascade GANs	–	Brain
PET (low-dose to full-dose)			
Xiang et al. [51]	Cascade CNNs	–	Brain
Wang et al. [46]	3D cGAN	–	Brain
Wang et al. [47]	3D cGAN with locality-adaptive module	–	Brain
Cross-modality synthesis			
MR to CT or CT to MR			
Nie et al. [36]	3D CNN-FCN	–	Pelvic
Han et al. [20]	U-net	–	Brain
Leynes et al. [31]	U-net	–	Pelvic
Liu et al. [32]	CNN based autoencoder	–	Brain
Chartsias et al. [7]	cycleGAN	MM-WHS[f]	Cardiac
Nie et al. [37]	Cascade GANs	ADNI[e]	Brain and pelvic
Emami et al. [17]	cGAN	–	Brain
Hiasa et al. [21]	cycleGAN with gradient loss	–	Musculoskeletal
Zhang et al. [59]	cycleGAN with segmentors	–	Cardiac
CT to PET or PET to CT			
Ben et al. [3]	FCN-cGAN	–	Liver
Bi et al. [4]	cGAN with tumor label input	–	Thorax
Armanious et al. [2]	cGAN with CasNet generator	–	Brain
MR to PET or PET to MR			
Choi et al. [13]	cGAN (pix2pix)	ADNI[e]	Brain
Wei et al. [49]	Cascade GANs	–	Brain
Cross-modality MR (T1, T2, FLAIR, and MRA)			
Van et al. [42]	Location-sensitive CNN	NAMIC[d]	Brain
Chartsias et al. [8]	CNN based encoder and decoder	ISLES2015[g] BRATS2015[h] IXI[i]	Brain
Dar et al. [16]	cGAN (pix2pix)	MIDAS [5] BRATS2015[h] IXI[i]	Brain
Olut et al. [38]	cGAN	IXI[i]	Brain
Mok et al. [35]	cGAN (two generators and four multi-scale discriminators)	BRATS2015[h]	Brain

(continued)

Table 1 (continued)

Publication	Method	Dataset	Organ
Yang et al. [52]	cGAN	BRATS2015[h]	Brain
Welander et al. [50]	cycleGAN and UNIT	Human Connectome[j]	Brain
Yu et al. [56]	3D cGAN	BRATS2015[h]	Brain
Yu et al. [57]	3D cGAN with edge map adversarial learning	BRATS2015[h] IXI[i]	Brain

[a] https://dcm.bmia.nl/ncia/login.jsf
[b] https://www.aapm.org/GrandChallenge/LowDoseCT/
[c] http://www.bic.mni.mcgill.ca/brainweb/
[d] http://hdl.handle.net/1926/1687
[e] www.adniinfo.org
[f] http://www.sdspeople.fudan.edu.cn/zhuangxiahai/0/mmwhs/
[g] http://www.isles-challenge.org/ISLES2015/
[h] https://sites.google.com/site/braintumorsegmentation/home/brats2015
[i] http://brain-development.org/ixi-dataset/
[j] https://ida.loni.usc.edu/login.jsp.

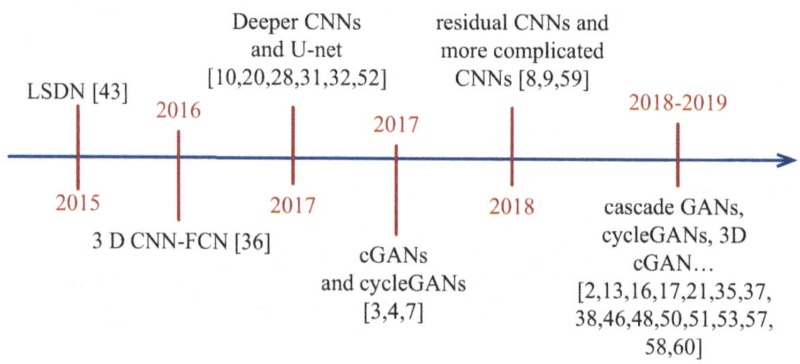

Fig. 1 The development of deep learning models for medical image synthesis

Deep Learning Models for Medical Image Synthesis

The state-of-the-art medical image synthesis methods usually use convolutional neural networks (CNN) [29]. With the delicate design, they can be exploited for whole-image or large-patch based synthesis to capture the implicit dependency among pixels/voxels in the same input during the end-to-end training. Among them, the most typical CNN architecture is U-net [40]. More recently, a number of CNN based generative adversarial networks (GANs) further improve the medical image synthesis results [55]. Therefore, this section will present the details of the typical conventional CNN model, i.e., U-net, and the basic GAN model in the research area of medical image synthesis.

Convolutional Neural Networks

The U-net model can extract the global contextual information from the source image and also reserve the spatially continuous details in the target image. As illustrated in Fig. 2, the original U-net model consists of the contracting and expanding paths. The number of convolutional layers in the contracting path is same to that in the expanding path. Between these two paths, multiple skip-connections are built to bridge them. With this structure, the U-net model can acquire the multi-depth information of the input source image. In addition, the gradient vanishing problem which is commonly shown during the training of deep learning models is mitigated, since the gradient of the deeper layers can be directly back-propagated to the shallower layers via the skip-connections.

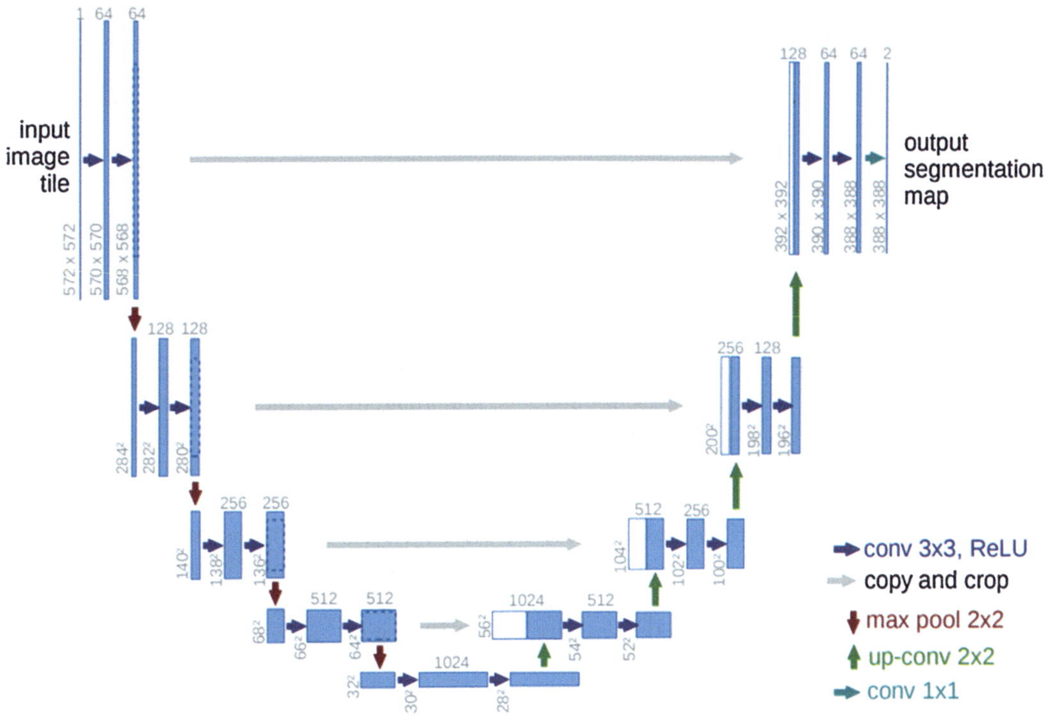

Fig. 2 Original 2D U-net model taken from [40]

This specific CNN architecture extracts the hierarchical visual clues from the input and is particularly suitable for the medical image synthesis tasks.

Generative Adversarial Networks

In 2014, the original GANs were first proposed for the generic image synthesis tasks [19]. Different from the common CNN based deep learning models, a GAN model consists of two agents, i.e., a generator G and a discriminator D, and is trained by the adversarial learning, as shown in Fig. 3. The original GAN model aims to learn a mapping from an input random noise to a target image that follows the distribution p_{data} of real images. In order to condition the GAN model on an input data of auxiliary information which could guide the mapping processing, conditional generative adversarial networks (cGANs) were then proposed [34]. When the input data is a source image $\mathbf{x} \sim p_{\text{data}}(\mathbf{x})$, the cGANs can be trained to synthesize its corresponding target image $\mathbf{y} \sim p_{\text{data}}(\mathbf{y})$ with the specific control from $\mathbf{x}$. This is a process of paired image-to-image synthesis.

Most existing GAN models for medical image synthesis [14, 16, 46, 57] follow a representative work pix2pix [25] and achieve very promising results. As a cGAN model, it synthesizes an image $G(\mathbf{x})$ from the given source image $\mathbf{x}$ to resemble the real target image $\mathbf{y}$ by its generator G. At the same time, its discriminator D is trained to differentiate between the synthesized image pair $(\mathbf{x}, G(\mathbf{x}))$ and the corresponding real image pair $(\mathbf{x}, \mathbf{y})$. Therefore, the synthesis performance can be improved through the adversarial competition between these two agents. The training loss of the generator G is formulated as follows:

$$\mathcal{L}^{G}_{\text{cGAN}} = \mathbb{E}_{\mathbf{x}\sim p_{\text{data}}(\mathbf{x})}[\log(1 - D(\mathbf{x}, G(\mathbf{x})))
+ \lambda_{l1}\mathbb{E}_{\mathbf{x},\mathbf{y}\sim p_{data}(\mathbf{x},\mathbf{y})}[\|y - G(\mathbf{x})\|_1], \quad (1)$$

where the symbol $\mathbb{E}$ denotes mathematical expectation, and $G(\cdot)$ and $D(\cdot)$ accordingly refer to the

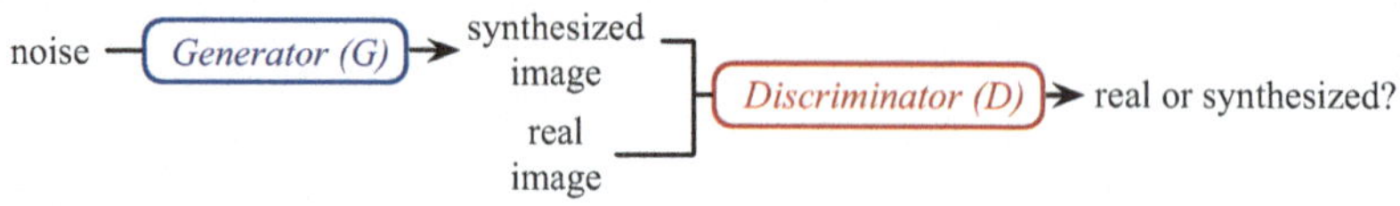

Fig. 3 Original GAN model

outputs of the generator and the discriminator. In Eq. (1), the generator G is trained to synthesize a realistic image which could fool the discriminator D via the first term. In its second term, the generator G tries to enforce the intensity similarity between the synthesized and real images through an L1-norm penalty on the pixel-wise intensity difference. The symbol λ_{l1} is a hyper-parameter to balance these two terms.

The loss function of the discriminator D is defined as follows:

$$\mathcal{L}_{\text{cGAN}}^{D} = -\mathbb{E}_{\mathbf{x},\mathbf{y}\sim p_{\text{data}}(\mathbf{x},\mathbf{y})}[\log D(\mathbf{x},\mathbf{y})] - \mathbb{E}_{\mathbf{x}\sim p_{\text{data}}(\mathbf{x})}[\log(1 - D(\mathbf{x}, G(\mathbf{x})))]. \tag{2}$$

In Eq. (2), different from the generator G, the discriminator D is trained to estimate the correct labels (0 or 1) for the synthesized or real image pairs. Thus, the adversarial competition between G and D conforms to a two-player min-max game.

In this cGAN model, the two sub-jobs of image generation and image discrimination are achieved together. Therefore, the final loss function integrates the above two objectives as follows:

$$\mathcal{L}_{\text{cGAN}} = \mathcal{L}_{\text{cGAN}}^{G} + \mathcal{L}_{\text{cGAN}}^{D}. \tag{3}$$

Both the generator and the discriminator in this cGAN model are CNN based to capture the powerful deep learning features. Specially, the generator has a U-net-like architecture to obtain the hierarchical contextual information from the input source images and then generate the better target images.

Within-Modality Synthesis

In this section, we present our two recent works for within-modality synthesis. These two works aim to synthesize high-quality positron emission tomography (PET) images to reduce the dose of radioactive tracer during the PET scanning. Since PET is widely exploited to visualize metabolism processes of human in clinics and research, it is important to get the clear PET images for patients. Before the PET image scanning, a full-dose radioactive tracer on a biologically active molecule is injected into the patient's body. During the scanning, the gamma rays which are emitted from the radioactive tracer in the body can be detected by the PET scanner. After that, the PET scanner analyzes the detected gamma rays of the full-dose tracer and constructs a high-quality three-dimensional (3D) PET image. However, the injected full-dose radioactive tracer brings up the risk of radioactive exposure and also raises the concerns about potential health hazards. As reported in "Biological Effects of Ionizing Radiation (BEIR VII),"[1] one full-dose radioactive tracer for every brain PET scan will improve the potential of lifetime cancer by 0.04%. When patients should take multiple times of PET scanning during their treatment, these risks will even accumulate, especially for the pediatric patients. To handle the radiation exposure issue, some researchers have lowered the injected dose of the tracer to the half of the full-dose, which inevitably decreases the quality of scanned PET images. The comparison of the full-dose PET image (F-PET) and the low-dose PET image (L-PET) is given in Fig. 4. Therefore, this high-quality PET image synthesis task aims to estimate the F-PET images from the given L-PET images.

[1] http://www.nap.edu/catalog/11340/health-risks-from-exposure-tolowlevels-of-ionizing-radiation.

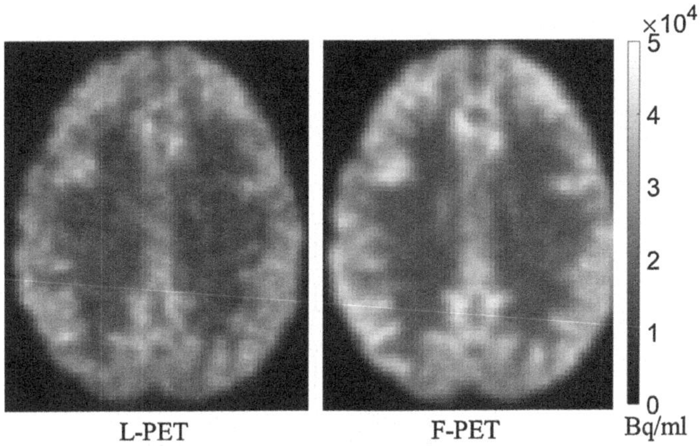

Fig. 4 Comparison between the low-dose PET (L-PET) image and the corresponding full-dose PET (F-PET). Taken from [47]

3D cGAN

Many cGAN models for medical image synthesis are 2D based [4,14,16,54]. When they are applied to the 3D medical imaging data, like PET and MRI, these methods first separate the 3D source image into axial slices and then separately map these source slices to the 2D target slices. After the concatenation of these synthesized 2D slices, a 3D target image can be reconstructed. Thus, the coronal and the sagittal slices of the reconstructed 3D target image are formed by the independently synthesized lines from the estimated axial planes. This will inevitably cause the loss of contextual information along the sagittal and coronal directions and the strong discontinuities in the final image. To mitigate this issue, our work in [46] proposes a 3D cGAN model to estimate the high-quality F-PET image from the L-PET image.

Framework

The framework of 3D cGAN model is illustrated in Fig. 5. Similar to the aforementioned cGAN model, this 3D cGAN consists of two agents: a 3D U-net-like generator G and a 3D CNN based discriminator D. The generator G processes a given L-PET image which is the source image and generates a synthesized F-PET image to approximate a real F-PET image. Simultaneously, the discriminator D is trained to distinguish between the synthesized F-PET-like image pair and the real F-PET image pair. The adversarial learning in the 3D cGAN follows the competition between two sub-tasks, i.e., the image generation of G and the image discrimination of D.

Experimental Results

As reported in [46], the 3D cGAN is evaluated on a real human brain dataset, which contains two categories: eight normal subjects and eight subjects diagnosed as mild cognitive impairment (MCI). Experiments are conducted in the widely used "Leave-One-Subject-Out cross-validation" strategy, i.e., in each experiment one subject is used as test data and the other 15 subjects are applied for training. To acquire sufficient 3D training data, 125 large image patches of size 64 × 64 × 64 are extracted from every original PET image of size 128 × 128 × 128 with the stride of 16. In the final synthesized 3D PET image, the overlapped regions are averaged from the estimated large patches. To evaluate the PET synthesis performance, peak signal-to-noise (PSNR) and normalized mean squared error (NMSE) are used.

To validate the effectiveness of the 3D model, 2D cGANs are compared with the 3D cGAN model. These 2D cGANs are separately trained with the 2D slices from the corresponding axial, coronal, and sagittal views. One visual example of synthesized results by the compared methods

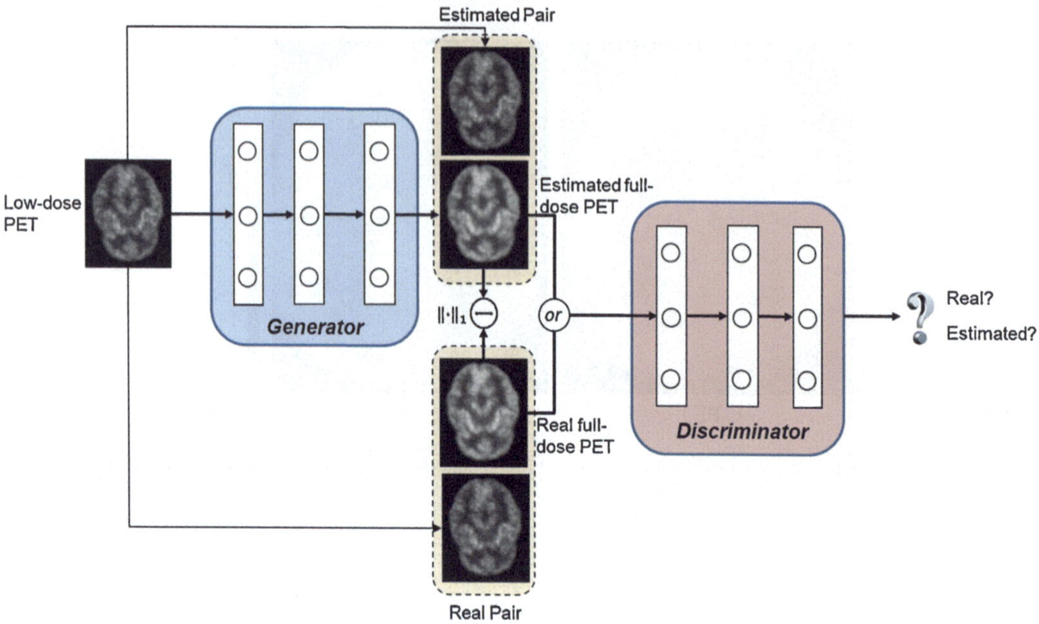

Fig. 5 Framework of training a 3D cGAN to estimate the full-dose PET image from low-dose counterpart. Taken from [46]

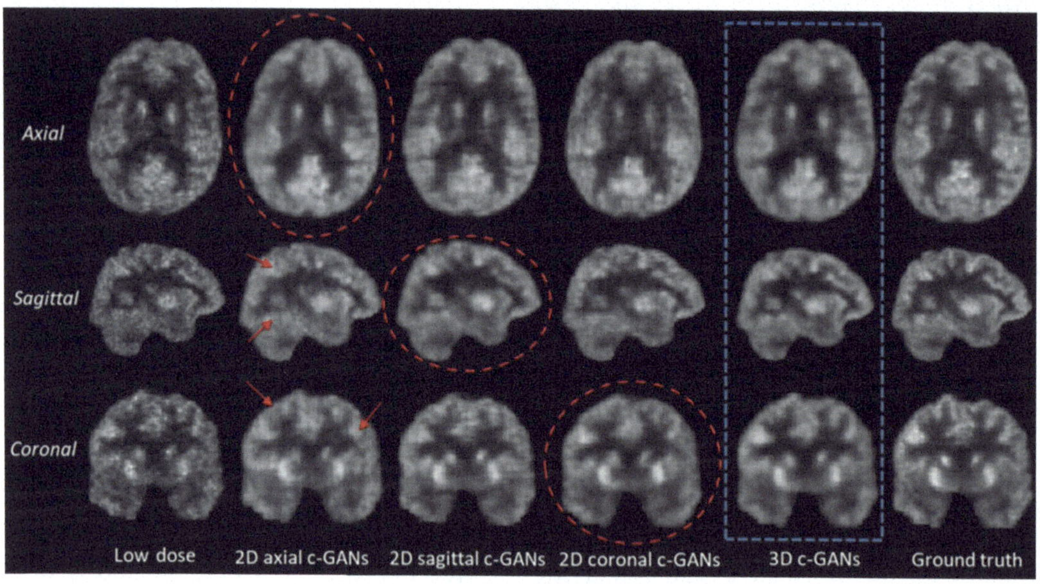

Fig. 6 Qualitative comparison between the results estimated by 2D cGANs and 3D cGANs. In the axial and coronal images, the left side of the image is the right side of the brain, and the right side of the image is the left side of the brain, taken from [46]

is given in Fig. 6. As shown, the results by 3D cGAN, which are presented in the blue block, have high visual quality in all three views. In contrast, these three 2D cGANs only produce good results in their corresponding trained views as indicated in the red circles, but get blurred synthesized views along the other two directions. The given example shows that the 2D cGANs

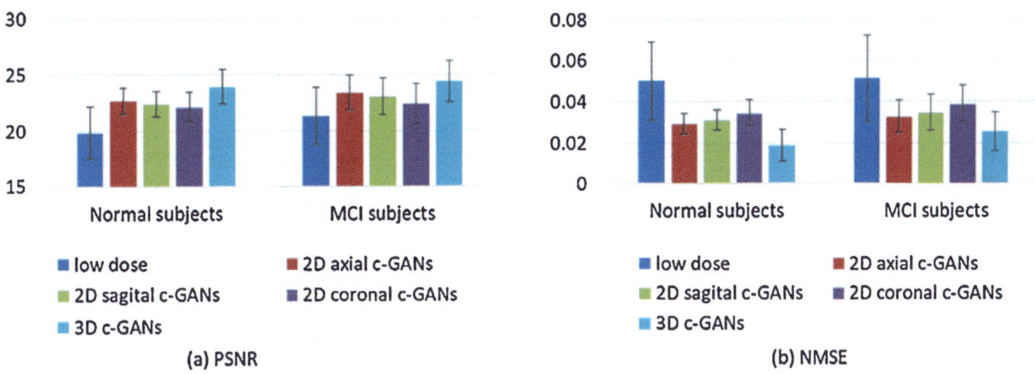

Fig. 7 Quantitative comparison between 2D cGANs and 3D cGANs, in terms of PSNR and NMSE, taken from [46]. Error bar indicates the standard deviation

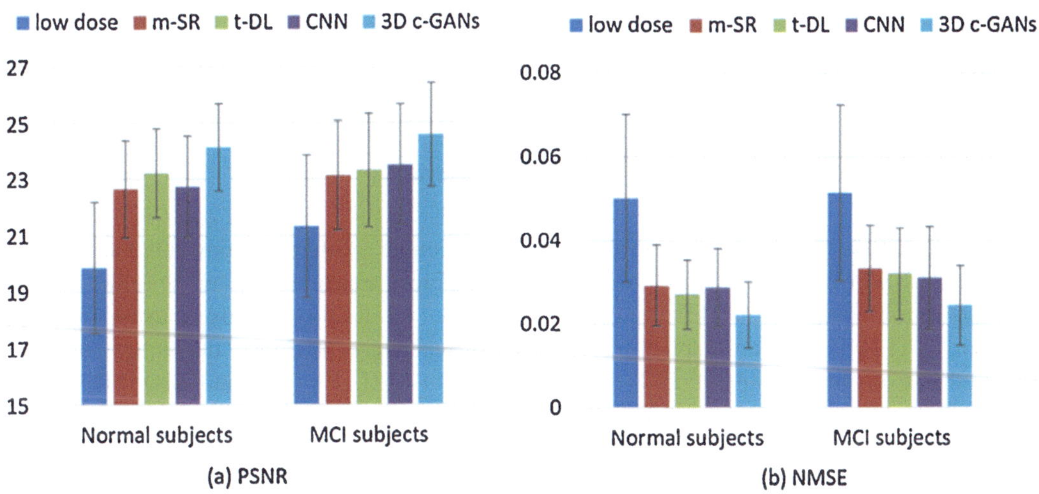

Fig. 8 Quantitative comparison between the existing PET estimation methods and the proposed method, in terms of PSNR and NMSE, taken from [46]. Error bar indicates the standard deviation

cause the discontinuous estimation across slices and lose the 3D structural information during the synthesis. The quantitative results of PSNR and NMSE are separately reported on the normal and the MCI data in Fig. 7. 3D cGAN achieves the best PSNR and NMSE results on both two categories of PET data, which consistently indicates that the 3D information captured in 3D cGAN can boost the synthesis.

Three state-of-the-art PET synthesis methods are compared with the 3D cGAN. They are (1) mapping based sparse representation (m-SR) [44], (2) semi-supervised tripled dictionary learning method (t-DL) [45], and (3) common CNN based method [51]. The quantitative comparison results are reported in Fig. 8. 3D cGAN performs best among all four methods in terms of both PSNR and NMSE, which demonstrates the superiority of the 3D cGAN in full-dose PET image synthesis.

Locality Adaptive Multi-Modality GANs

Recent research reports that using multiple modalities, like PET and MRI, benefits the medical image quality enhancement [23]. In addition, different from PET, scanning MRI would not raise the risks of radioactive exposure

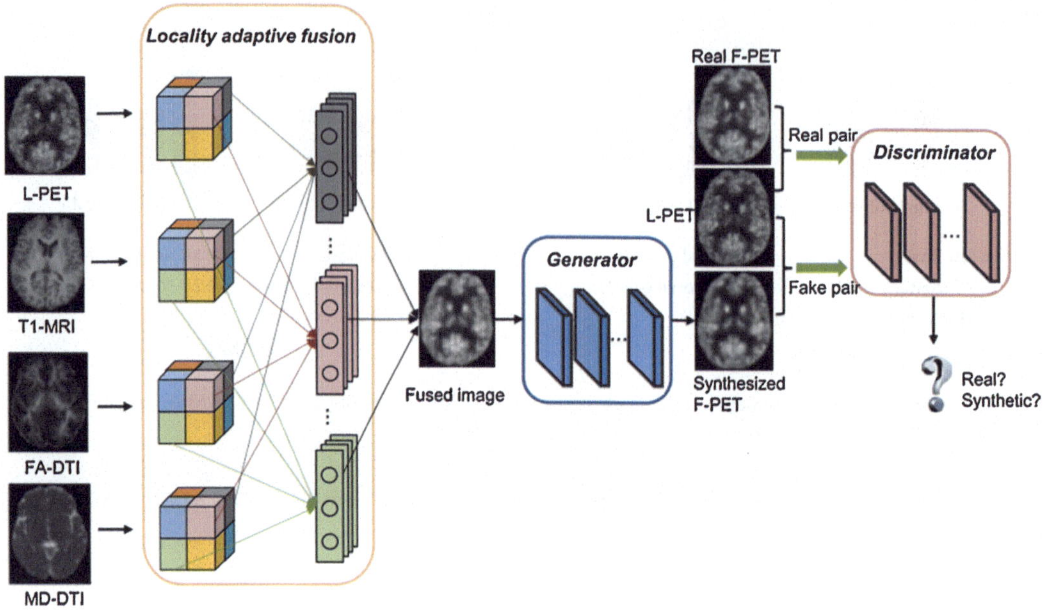

Fig. 9 Overview of locality adaptive multi-modality GANs, taken from [47]

for patients. Thus, medical images of T1-weighted MRI (T1-MRI), fractional anisotropy diffusion tensor image (FA-DTI), and mean diffusivity DTI (MD-DTI) are applied to assist the synthesis of high-quality F-PET images from L-PET images. Traditionally, the image convolution in CNN based GANs is performed on these multiple images (input channels) in a global manner. That is to say, the common multi-channel based GANs apply the same convolution filter to all image locations of each input modality for producing the feature maps which will be combined in deeper layers. As a result, these multi-channel models would not consider the location-varying contributions from the various input modalities. To handle this issue, locality adaptive multi-modality GANs (LA-GANs) are proposed in [47] for PET image synthesis.

Framework

The LA-GANs model includes three modules: (1) the locality-adaptive fusion network, (2) the generator network, and (3) the discriminator network, as illustrated in Fig. 9. The newly added locality-adaptive fusion network processes L-PET, T1-MRI, FA-DTI, and MD-DTI images as input channels and estimates a fused image by learning different convolutional kernels at different image locations. Specifically, the module of locality-adaptive fusion network first separately partitions the entire input L-PET, T1-MRI, FA-DTI, and MD-DTI images into N non-overlapped small patches which are accordingly denoted by P_i^L, P_i^{T1}, P_i^{FA}, and P_i^{MD} $(i = 1, \ldots, N)$. These small patches from different locations are indicated by different colors in Fig. 9. After that, the patches at the same location from the four input modalities are separately convolved by four different $1 \times 1 \times 1$ filters with parameters ω_i^L, ω_i^{T1}, ω_i^{FA}, and ω_i^{MD}, respectively. Through this locality-adaptive convolution, a fused patch P_i^C can be calculated as follows:

$$\begin{aligned} P_i^C &= \omega_i^L P_i^L + \omega_i^{\mathrm{T1}} P_i^{\mathrm{T1}} + \omega_i^{\mathrm{FA}} P_i^{\mathrm{FA}} + \omega_i^{\mathrm{MD}} P_i^{\mathrm{MD}}, \\ & s.t.\ \omega_i^L + \omega_i^{\mathrm{T1}} + \omega_i^{\mathrm{FA}} + \omega_i^{\mathrm{MD}} = 1, \\ & \omega_i^L, \omega_i^{\mathrm{T1}}, \omega_i^{\mathrm{FA}}, \omega_i^{\mathrm{MD}} > 0, i = 1, \ldots, N. \end{aligned} \tag{4}$$

Therefore, N groups of different convolution filters for the $N * 4$ small patches at N locations from four modalities can be learned.

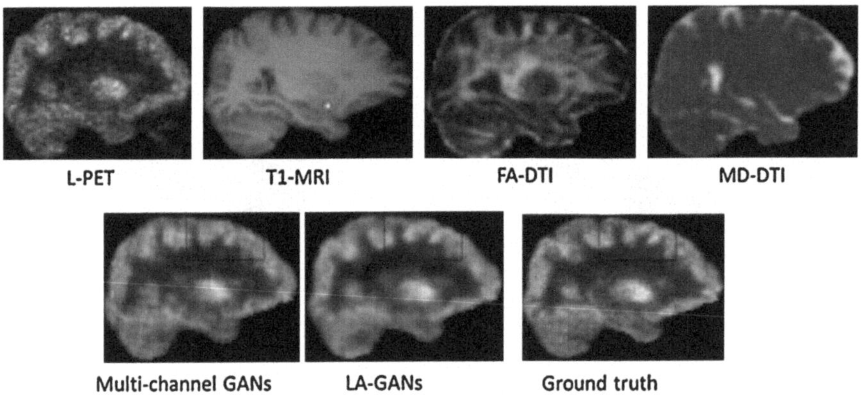

Fig. 10 Visual comparison with multi-channel GANs method, taken from [47]

Table 2 Quantitative comparison with the multi-channel GANs method on normal and MCI subjects

	Methods	PSNR		SSIM	
		Mean (std.)	Med.	Mean (std.)	Med.
Normal subjects	L-PET	19.88 (2.34)	20.68	0.9790 (0.0074)	0.980
	Multi-channel	24.36 (1.93)	24.78	0.9810 (0.0065)	0.983
	LA-GANs	**24.61 (1.79)**	**25.32**	**0.9860 (0.0053)**	**0.987**
MCI subjects	L-PET	21.33 (2.53)	21.62	0.9760 (0.0102)	0.979
	Multi-channel	24.99 (2.03)	25.36	0.9795 (0.0098)	0.982
	LA-GANs	**25.19 (1.98)**	**25.54**	**0.9843 (0.0097)**	**0.988**

Mean (standard deviation), Median. The paired t-test of PSNR shows that our improvement against the multi-channel one is statistically significant with $p < 0.05$ ($p = 0.048$ for NC subjects and $p = 0.016$ for MCI subjects). For SSIM, our method also presents the significant improvement, with p-value 0.051 for NC subjects and 0.037 for MCI subjects, respectively

After the above locality-adaptive fusion, the final fused image is applied as the input of the generator to generate F-PET-like images. The generator and the discriminator in our LA-GANs work similarly to those in the aforementioned 3D cGAN.

Experimental Results

To evaluate the effectiveness of the newly added locality-adaptive fusion network module, the common multi-channel GANs model is compared with the LA-GANs. Figure 10 gives an example of visual results obtained by these two methods. We can observe that the LA-GANs model synthesizes the F-PET-like image with less artifacts than the compared multi-channel model, which are clearly indicated by red rectangles. The quantitative results of these two methods are reported in Table 2 via the evaluation measures of PSNR and structural similarity index (SSIM) [48]. The top part gives the results on normal subjects, and the bottom part reports the results on MCI subjects. These results show the superiority of the locality-adaptive fusion network module over the common multi-channel processing, in terms of both PSNR and SSIM. In addition, through conducting the paired t-test, all these improvements are statistically significant at the significance level of 0.05. Both the visual and quantitative results demonstrate the effectiveness of locality-adaptive fusion network in cGAN models for the full-dose PET synthesis task.

Moreover, the LA-GANs model is compared with four state-of-the-art methods, i.e., (1) mapping based sparse representation method (m-SR) [44], (2) tripled dictionary learning method (t-DL) [45], (3) multi-level CCA method (m-CCA) [1], and (4) auto-context CNN method

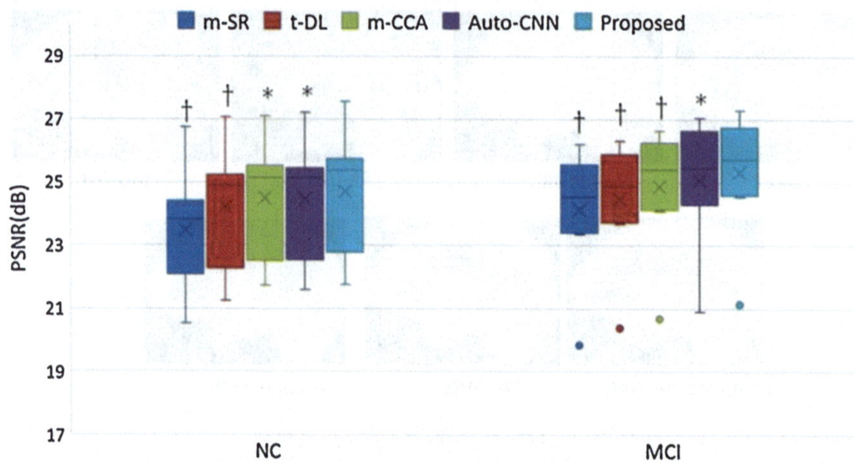

Fig. 11 Qualitative comparison with the state-of-the-art PET estimation methods in terms of PSNR, taken from [47]. † indicates $p < 0.01$ in the paired t-test while ⋆ means $p < 0.05$

(auto-CNN) [51], as shown in Fig. 11. The highest PSNR values of the LA-GANs indicate that it has the best performance among all the compared methods.

Cross-Modality Synthesis

Our two recent cross-modality synthesis works are presented in this section. They are utilized in cross-modality MR image synthesis that aims to better visualize the scanned body parts from diverse imaging perspectives and facilitate the following clinical applications, e.g., tumor segmentation. When setting different scanning parameters, MRI can generate multiple-modality images (e.g., T1-weighted, T2-weighted, and FLAIR) which show the diverse contrasts of soft tissues. Since each modality image provides the unique visual representation of scanned body parts, these multiple modalities are usually studied together in the subsequent analysis for disease diagnosis [15] and treatment planning [30]. However, due to the potential of modality loss in clinics, the quality of the analysis will be adversely affected. Therefore, cross-modality MR image synthesis is highly desirable to synthesize the unknown target-modality MR images from the given source-modality images [26, 53].

3D cGAN with Subject-Specific Local Adaptive Fusion

Our work in [56] proposes a 3D cGAN based cross-modality MR image synthesis method to boost brain tumor segmentation performance. Compared with the single synthesis task, this is more challenging and requires the higher quality of synthesized images because of two main reasons. First, due to the arbitrary location and appearance of brain tumor, the pathology involved MR images raise the difficulty of synthesis in contrast to the healthy subject images. Second, the source-modality images may lack some important pathology-related information which can be seen in the target-modality. For example, as shown in Fig. 12, the diffuse changes around tumor parts are only observed in the FLAIR image. Thus, [56] presents an additional approach to the 3D cGAN, which is called subject-specific local adaptive fusion. This fusion approach aims to polish the local details in the synthesized target-modality-like images from the 3D cGAN through a linear combination of the real target-modality images among the training set for approximation. During the combination, the combination weights are estimated from the synthesized target-modality-like images which are the outputs of the 3D cGAN model. In this way, this local and adaptive approach can improve

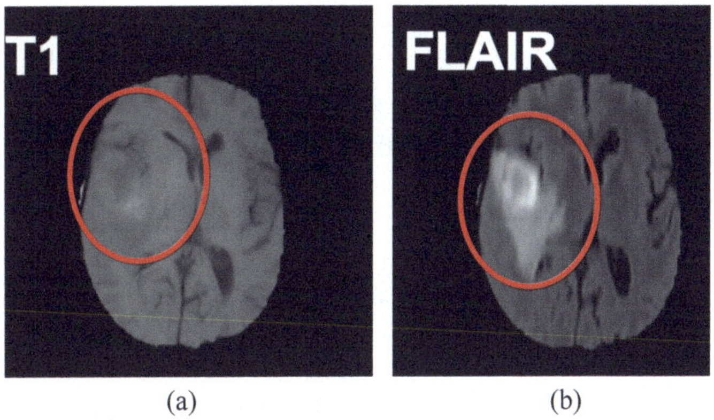

Fig. 12 A brain T1 image (**a**) and the corresponding FLAIR image (**b**)

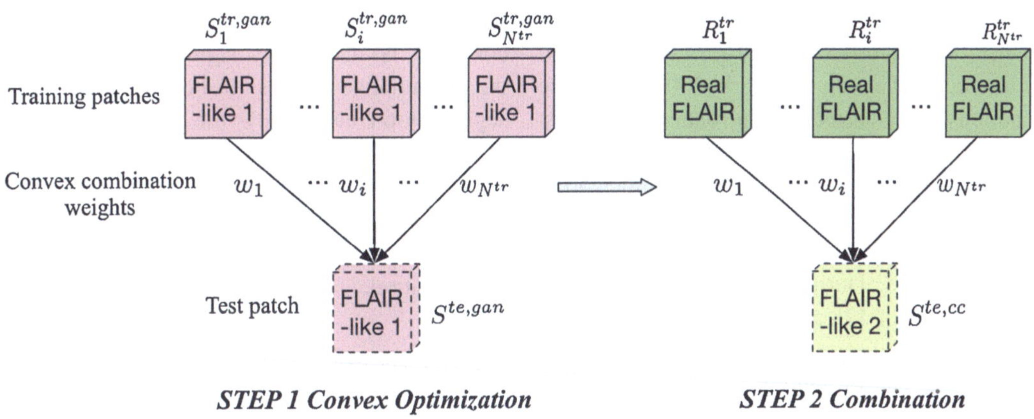

Fig. 13 Framework of subject-specific local adaptive fusion

the quality of synthesized images and further raise the segmentation performance.

Framework

The framework of this subject-specific local adaptive fusion is illustrated in Fig. 13. Here, we take T1-to-FLAIR synthesis task as an example, and the synthesized FLAIR-like image from the 3D cGAN and the final fused FLAIR-like image are called FLAIR-like-1 image and FLAIR-like-2 image, respectively. Before this fusion, for each test subject that only has its real T1 image, its corresponding FLAIR-like-1 image is partitioned into non-overlapped small patches of size 16 × 16 × 16. Each FLAIR-like-1 patch $S^{\text{te,gan}}$ of this test subject is approximated by the convex combination of the patches $S_1^{\text{tr,gan}}, S_2^{\text{tr,gan}} \ldots, S_{N_{\text{tr}}}^{\text{tr,gan}}$ at the same location from the FLAIR-like-1 images of training subjects. The symbol N_{tr} denotes the number of all training subjects. This approximation is achieved through the following optimization:

$$\min_{w} \|\sum_{i=1}^{N_{\text{tr}}} w_i S_i^{\text{tr,gan}} - S^{\text{te,gan}}\|_2^2 \quad s.t. \sum w_i = 1, w_i \geq 0. \tag{5}$$

Therefore, the combination weights w_i $(i = 1, \ldots, N_{\text{tr}})$ are learned via Eq. (5). Since the FLAIR-like-1 image is generated to resemble the corresponding real image, these combination weights could be further applied to linearly combine the real FLAIR training patches $R_1^{\text{tr}}, R_2^{\text{tr}} \ldots, R_{N_{\text{tr}}}^{\text{tr}}$ at the same location to polish the final FLAIR-like-2 patch $S^{te,cc}$ through the following equation:

$$S^{te,cc} = \sum_{i=1}^{N_{tr}} w_i R_i^{tr}. \quad (6)$$

In this way, a better polished target-modality-like image is estimated and used together with its corresponding real source-modality image in the subsequent brain tumor segmentation.

Experimental Results

This work is evaluated on BRATS2015 dataset [33], which includes 274 subjects of four modality images, i.e., T1, T1C, T2, and FLAIR, with size of 240 × 240 × 155, and additional brain tumor labels. In this work, 230 subjects are randomly selected as training data and the rest 44 subjects are test data for the T1-to-FLAIR synthesis task. The brain tumor segmentation model from [27] is utilized to evaluate the segmentation performance of the synthesized FLAIR images. During the synthesis task, 3D large patches of 128 × 128 × 128 are extracted from images to increase the training samples for 3D cGAN. The evaluation measures of PSNR and NMSE on the synthesized whole brains and tumor regions are utilized.

To study the contribution of the subject-specific local adaptive fusion, it (i.e.,3D cGAN (128)+local adaptive fusion) is compared with another two methods for ablation study. They are: (1) 3D cGAN trained on large patches (128^3) and (2) local non-linear mapping (3D cGAN on patches with the size of 32^3) applied after the method (1). The synthesis results are reported in Table 3. As shown, the 3D cGAN (128)+local adaptive fusion outperforms the other two methods, demonstrating the effectiveness of subject-specific local adaptive fusion in T1-to-FLAIR image synthesis task. The results also indicate that using the linear combination in the local adaptive fusion can obtain better results than the local non-linear mapping of 3D cGAN (32). Table 4 gives the segmentation results on whole tumor parts and tumor core regions by the above three methods, which consistently indicates the better performance of the 3D cGAN (128)+local adaptive fusion approach. The results by using the single modality of T1 are also reported. The paired t-test result verifies that the improvement on the tumor core part is statistically significant. Therefore, the quantitative results of both synthesis and segmentation tasks show the advantage of 3D cGAN (128)+local adaptive fusion approach in synthesizing FLAIR images from T1, and the benefits of using the synthesized FLAIR images to improve the T1-based brain tumor segmentation.

Edge-Aware GANs

The aforementioned cGAN models enforce the pixel/voxel-wise intensity similarity between the real and the synthesized images through using an L1-norm penalty during training. However, the structure of image content, like the textural information in MRI [6], is not sufficiently captured by these models. The edge information in an image provides the details about the textural structure of image content through capturing the local intensity changes and the boundaries

Table 3 Quantitative evaluation results of the synthesized images

Methods	Synthesis quality (PSNR/NMSE%)	
	Whole brain	Tumor
3D cGAN (128)	20.45/25.08	19.13/12.68
3D cGAN (128)+3D cGAN (32)	19.94/24.99	18.73/13.45
3D cGAN (128) + local adaptive fusion	**20.68/22.67**	**19.27/11.86**

Values with underline indicate they are statistically significantly different from 3D cGAN (128)+local adaptive fusion, according to a two-sided, paired t-test (solid line $p < 0.05$). t-Test values are given as follows: (1) proposed method over 3D cGAN: (a) whole brain ($1.17e-1/4.59e-2$); (b) tumor ($1.17e-2/8.13e-4$). (2) Proposed method over 3D cGAN(128) + (32): (a) whole brain ($8.75e-7/5.59e-4$); (b) tumor ($2.1e-3/6.12e-5$)

Table 4 Quantitative evaluation results of segmentation

Methods	Segmentation (dice ratio%)	
	Whole tumor	Core
3D cGAN (128)	66.35	72.09
3D cGAN (128)+3D cGAN (32)	66.61	72.14
T1	67.18	63.00
T1+real FLAIR (ideal scenario)	*82.17*	*85.49*
3D cGAN (128) + local adaptive fusion	**68.23**	**72.28**

Values with underline indicate they are statistically significantly different from 3D cGAN (128)+local adaptive fusion, according to a two-sided, paired t-test (solid line $p < 0.05$, dotted line $p < 0.1$). t-Test values are given as follows: (1) proposed method over 3D cGAN: (a) whole tumor (0.0643); (b) core (0.886). (2) Proposed method over 3D cGAN(128) + (32): (a) whole tumor (0.0672); (b) tumor (0.912). (3) Proposed method over T1: (a) whole tumor (0.262); (b) tumor ($9.44e - 5$)

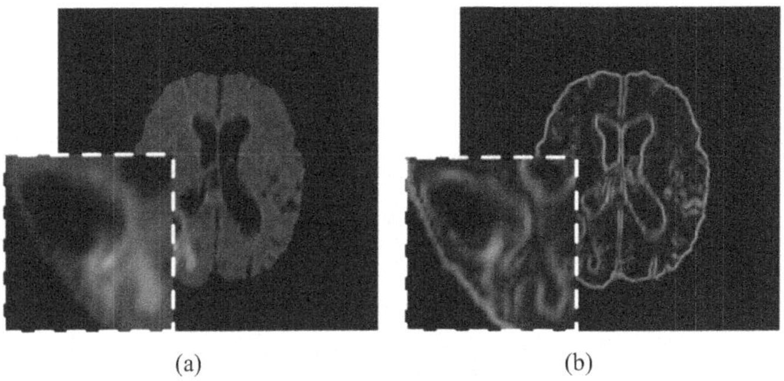

Fig. 14 A brain FLAIR image (**a**), and its corresponding edge map (**b**) after the 3D Sobel edge detection, taken from [57]. The contour of abnormal tissues can be depicted clearer by the edge map, which is shown as the zoomed regions

between different tissues. Thus, maintaining the edges during the synthesis can help to sharpen the synthesized target-modality MR images. Especially, for a pathology involved MR image, the edge details benefit to distinguish between the normal and the abnormal tissues, which is important to depict the contour of the arbitrary pathological regions. For example, Fig. 14 shows that the zoomed gliomas tumor is very clear in the edge map of a brain MR image. Therefore, our work in [57] proposes new cGAN models to enforce edge preservation for cross-modality MR image synthesis. This work adds an extra constraint to 3D cGAN models to realize edge-aware generative adversarial networks (Ea-GANs) by ensuring the similarity of the edge maps extracted from the real and the synthesized images during training. These edge maps are calculated via the commonly applied Sobel filters as shown in Fig. 15. These three $3 \times 3 \times 3$ Sobel filters, i.e., F_i, F_j, and F_k, are applied to convolve a given image A to produce its three edge maps which correspond to the intensity gradients along i, j, and k directions, respectively. After that, a final edge map $S(A)$ of A is obtained by merging the three-direction edge maps through the following equation:

$$S(A) = \sqrt{(F_i * A)^2 + (F_j * A)^2 + (F_k * A)^2}, \quad (7)$$

where $*$ denotes the convolution operation.

Framework

As shown in Fig. 16, [57] proposes two different frameworks, i.e., a generator-induced Ea-GAN (gEa-GAN) and a discriminator-induced Ea-GAN (dEa-GAN), according to the different strategies of using the edge maps. Both of these two Ea-GANs are composed of three modules: (1) a generator G, (2) a discriminator D, and (3) a Sobel edge detector S.

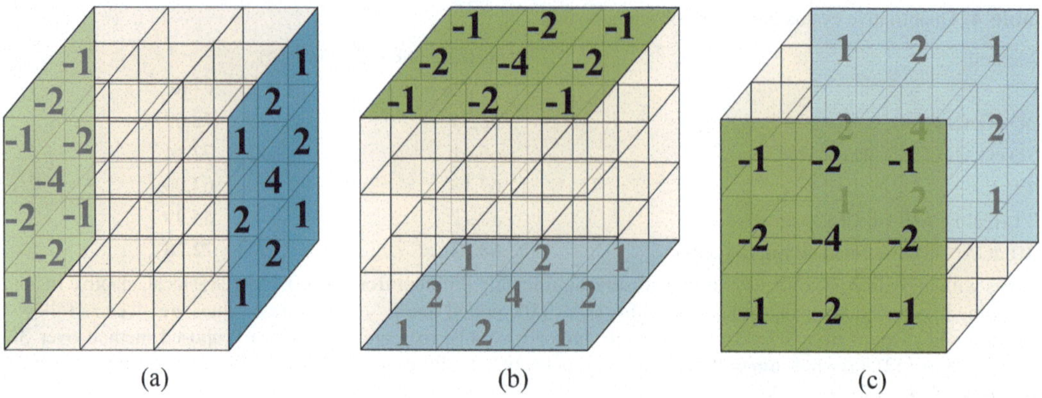

Fig. 15 Three-dimensional Sobel operator, taken from [57]. (**a**) F_i. (**b**) F_j. (**c**) F_k

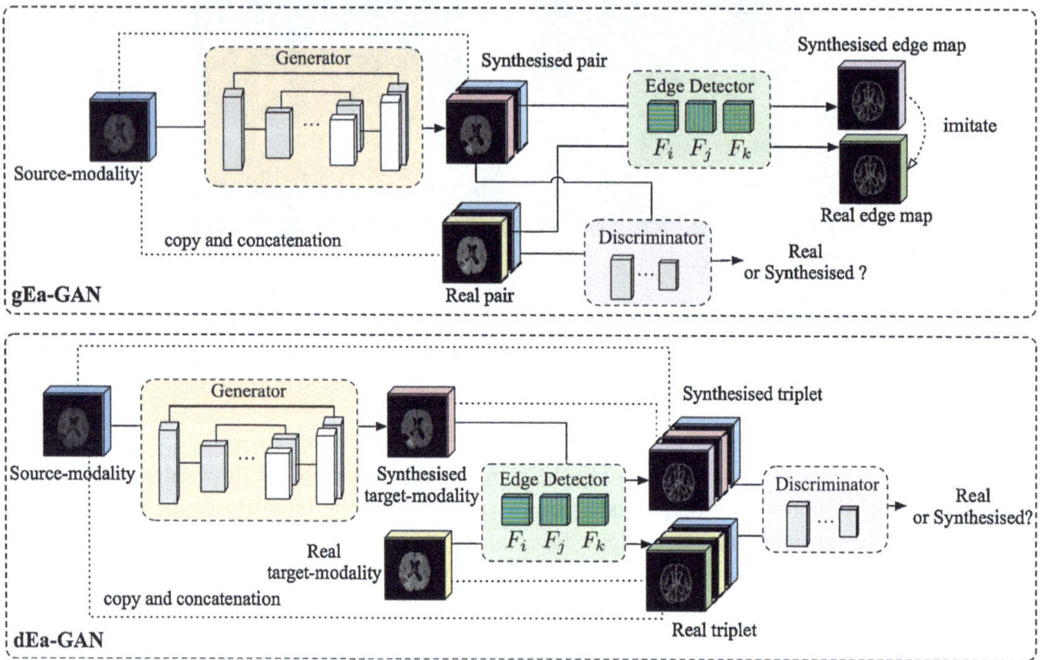

Fig. 16 Frameworks of Ea-GANs, taken from [57]

For the gEa-GAN model, when given a source-modality image $\mathbf{x}$ and its target-modality counterpart $\mathbf{y}$ as the groundtruth for the cross-modality MR image synthesis task, its generator G tries to synthesize target-modality-like images $G(\mathbf{x})$ that can be misclassified by its discriminator D through the adversarial learning. The L1-norm penalty through G is applied to ensure the voxel-wise intensity similarity between the real and the synthesized images, similar to the 3D cGAN model. Additionally, another L1-norm penalty is used to discourage the difference between their corresponding Sobel edge maps which are extracted from S during the training of gEa-GAN. Therefore, the loss function of the generator G is formulated as follows:

$$
\begin{aligned}
\mathcal{L}^{G}_{\text{gEa-GAN}} = & \ \mathbb{E}_{\mathbf{x}\sim p_{\text{data}}(\mathbf{x})}[\log(1 - D(\mathbf{x}, G(\mathbf{x})))] \\
& + \lambda_{l1}\mathbb{E}_{\mathbf{x},\mathbf{y}\sim p_{\text{data}}(\mathbf{x},\mathbf{y})}[\|y - G(\mathbf{x})\|_1] \\
& + \lambda_{\text{edge}}\mathbb{E}_{\mathbf{x},\mathbf{y}\sim p_{\text{data}}(\mathbf{x},\mathbf{y})} \\
& [\|S(y) - S(G(\mathbf{x}))\|_1],
\end{aligned}
\tag{8}
$$

where the hyper-parameters, λ_{l1} and λ_{edge}, are used to balance the three terms in Eq. (8).

Similar to the case of 3D cGAN model, the loss function of its discriminator D is defined as follows:

$$\mathcal{L}^{D}_{\text{gEa-GAN}} = -\mathbb{E}_{\mathbf{x},y\sim p_{\text{data}}(\mathbf{x},\mathbf{y})}[\log D(\mathbf{x},\mathbf{y})] - \mathbb{E}_{\mathbf{x}\sim p_{\text{data}}(\mathbf{x})}[\log(1-D(\mathbf{x},G(\mathbf{x})))]. \tag{9}$$

Thus, the final objective function of gEa-GAN integrates the above two loss functions of the generator and the discriminator as follows:

$$\mathcal{L}_{\text{gEa-GAN}} = \mathcal{L}^{G}_{\text{gEa-GAN}} + \mathcal{L}^{D}_{\text{gEa-GAN}}. \tag{10}$$

Different from the gEa-GAN that maintains the edge similarity only by its generator during training, the dEa-GAN model additionally brings the edge information into the adversarial learning between its generator and discriminator. In this way, its discriminator could also perceive the edge details of the synthesized images and further benefit the synthesis processing. Specifically, the edge maps of the real and the synthesized target-modality images are correspondingly concatenated with the real and the synthesized image pairs as the real triplet ($\mathbf{x}$, y, $S(\mathbf{y})$) and the synthesized triplet ($\mathbf{x}$, $G(\mathbf{x})$, and $S(G(\mathbf{x}))$. The discriminator D of dEa-GAN tries to distinguish between these two kinds of triplets, and this in turn enforces its generator G to estimate the better edge details for synthesis.

For dEa-GAN, its generator G is also trained by the adversarial loss, the voxel-wise intensity difference loss, and the edge difference loss for synthesis, following the designed objective:

$$\begin{aligned}\mathcal{L}^{G}_{\text{dEa-GAN}} &= \mathbb{E}_{\mathbf{x}\sim p_{\text{data}}(\mathbf{x})}[\log(1-D(\mathbf{x},G(\mathbf{x}),S(G(\mathbf{x}))))\\ &+ \lambda_{l1}\mathbb{E}_{\mathbf{x},\mathbf{y}\sim p_{\text{data}}(\mathbf{x},\mathbf{y})}[\|y-G(\mathbf{x})\|_1]\\ &+ \lambda_{\text{edge}}\mathbb{E}_{\mathbf{x},\mathbf{y}\sim p_{\text{data}}(\mathbf{x},\mathbf{y})}[\|S(\mathbf{y})-S(G(\mathbf{x}))\|_1].\end{aligned} \tag{11}$$

Different from the gEa-GAN model, the edge map $S(G(\mathbf{x}))$ in dEa-GAN is implicitly utilized in the first term of Eq. (11) through calculating the loss error of the outputs from its discriminator D.

The objective function of the discriminator D is accordingly designed as:

$$\begin{aligned}\mathcal{L}^{D}_{\text{dEa-GAN}} &= -\mathbb{E}_{\mathbf{x},\mathbf{y}\sim p_{\text{data}}(\mathbf{x},\mathbf{y})}[\log D(\mathbf{x},\mathbf{y},S(\mathbf{y}))]\\ &- \mathbb{E}_{\mathbf{x}\sim p_{\text{data}}(\mathbf{x})}\\ &[\log(1-D(\mathbf{x},G(\mathbf{x}),S(G(\mathbf{x}))))].\end{aligned} \tag{12}$$

Finally, the objective for training the entire dEa-GAN model is

$$\mathcal{L}_{\text{dEa-GAN}} = \mathcal{L}^{G}_{\text{dEa-GAN}} + \mathcal{L}^{D}_{\text{dEa-GAN}}. \tag{13}$$

Experimental Results

The Ea-GANs are evaluated on BRATS2015 dataset by the way of 5-fold cross validation. They are compared with five methods: (1) handcrafted feature used replica [26], (2) common CNN based multimodal [8], (3) 2D cGAN based pix2pix [25], (4) 3D cGAN, and (5) gradient loss utilized gradient cGAN. The evaluation measures of PSNR, NMSE, and SSIM are separately applied on the synthesized whole images including the background and the brain part. Two synthesis tasks, T1-to-FLAIR and T1-to-T2, are conducted to show the performance of Ea-GANs for cross-modality MR image synthesis. Their quantitative and visual results are presented in Table 5, Fig. 17, Table 6, and Fig. 18.

When comparing two Ea-GANs with the 3D cGAN model through the given quantitative results, the Ea-GANs produce higher-quality images than 3D cGAN with the significant improvements of PSNR from 29.26dB (3D cGAN) to 30.11dB (dEa-GAN), SSIM from 0.958 (3D cGAN) to 0.963 (dEa-GAN), and NMSE from 0.119 (3D cGAN) to 0.105 (dEa-GAN), respectively, in the T1-to-FLAIR task. Similarly, these two Ea-GANs also outperform the 3D cGAN in the T2 image synthesis task. These results demonstrate the effectiveness of preserving edge information in the synthesized images. Additionally, the dEa-GAN model performs better than the gEa-GAN model in both of two synthesis

Table 5 Quantitative evaluation results of the synthesized FLAIR-like from T1 on the BRATS2015 dataset (mean ± standard deviation)

Methods	PSNR	NMSE	SSIM
Replica [26]	27.17 ± 2.60	0.171 ± 0.267	0.939 ± 0.013
Multimodal [8]	27.26 ± 2.82	0.184 ± 0.284	0.950 ± 0.014
Pix2pix [25]	27.46 ± 2.55	0.144 ± 0.189	0.940 ± 0.015
3D cGAN	29.26 ± 3.21	0.119 ± 0.205	0.958 ± 0.016
Gradient cGAN	29.38 ± 3.25	0.116 ± 0.204	0.960 ± 0.017
gEa-GAN	29.55 ± 3.24	0.115 ± 0.199	0.960 ± 0.017
dEa-GAN	**30.11 ± 3.22**	**0.105 ± 0.174**	**0.963 ± 0.016**

The paired t-test is conducted between dEa-GAN and a compared method at the significance level of 0.05. When the improvement of dEa-GAN over the method is statistically significant, the result of that compared method will be underlined. *t*-Test values of proposed dEa-GAN over the following methods: (a) Replica: $7.96e-53$; $3.94e-13$; $3.67e-106$. (b) Multimodal: $9.10e-39$; $1.75e-12$; $2.11e-41$. (c) Pix2pix: $7.72e-67$; $8.50e-21$; $8.71e-131$. (d) 3D cGAN: $1.05e-63$; $6.62e-6$; $6.84e-42$. (e) Gradient cGAN: $5.29e-30$; $1.08e-2$; $1.80e-22$. (f) gEa-GAN: $3.41e-25$; $9.15e-4$; $1.67e-15$

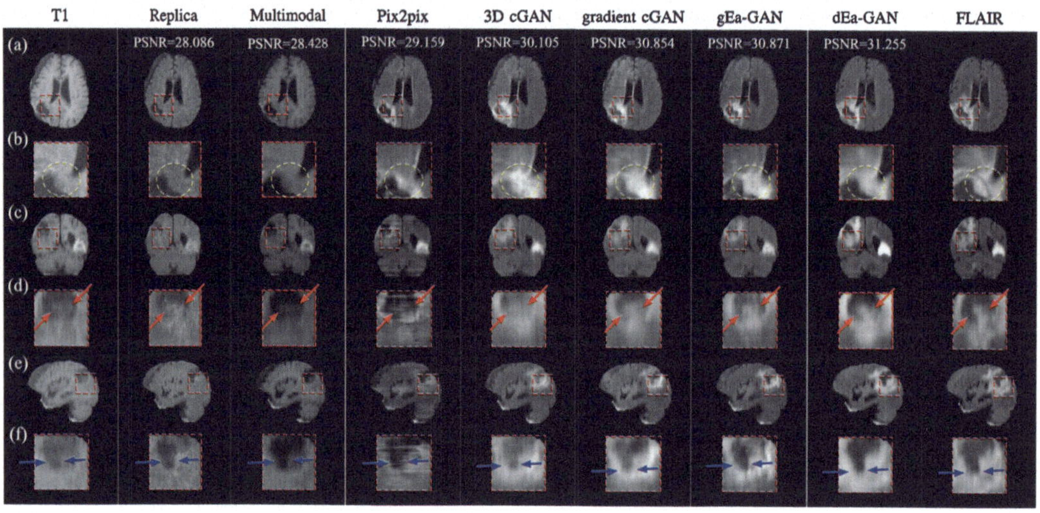

Fig. 17 Visual comparison of the synthesized FLAIR images between Ea-GANs and other state-of-the-art methods taken from [57]: (**a**) axial slices, (**b**) zoomed parts of axial slices, (**c**) coronal slices, (**d**) zoomed parts of coronal slices, and (**e**) sagittal slices, (**f**) zoomed parts of sagittal slices

tasks, showing the necessity of bringing the edge information into the training of the discriminator. Furthermore, the superiority of the Ea-GANs is consistently shown when compared with the other four state-of-the-art methods in terms of all the three measures. When looking into the visual comparison examples in Figs. 17 and 18, from all the three views, it can be seen that the Ea-GANs synthesize sharper edges and more local details than the compared methods as indicated. Therefore, both the quantitative and visual results demonstrate that the Ea-GANs synthesize better MR images by using edge maps via two different strategies in cGAN models than the compared methods.

Conclusion

In this chapter, we focus on presenting deep learning approaches for medical image synthesis. Through the experimental results in our four

Table 6 Quantitative evaluation results of the synthesized T2-like images from T1 on the BRATS2015 dataset (mean ± standard deviation)

Methods	PSNR	NMSE	SSIM
Replica [26]	26.92 ± 2.36	0.158 ± 0.324	0.946 ± 0.015
Multimodal [8]	27.31 ± 2.39	0.140 ± 0.229	0.951 ± 0.016
Pix2pix [25]	28.12 ± 2.45	0.110 ± 0.220	0.953 ± 0.014
3D cGAN	29.34 ± 3.23	0.095 ± 0.199	0.964 ± 0.017
Gradient cGAN	29.43 ± 3.28	0.097 ± 0.210	0.966 ± 0.017
Proposed gEa-GAN	29.58 ± 3.29	0.093 ± 0.218	0.966 ± 0.018
Proposed dEa-GAN	**29.98 ± 3.37**	**0.088 ± 0.223**	**0.967 ± 0.016**

The paired t-test is conducted between dEa-GAN and a compared method at the significance level of 0.05. When the improvement of dEa-GAN over the method is statistically significant, the result of that compared method will be underlined. t-Test values of proposed dEa-GAN over the following methods: (a) Replica: $4.03e-41$; $4.21e-7$; $1.85e-72$. (b) Multimodal: $4.25e-48$; $4.32e-23$; $1.49e-78$. (c) Pix2pix: $1.90e-42$; $2.19e-9$; $2.44e-106$. (d) 3D cGAN: $4.25e-30$; $1.54e-4$; $3.43e-40$. (e) Gradient cGAN: $2.24e-33$; $3.77e-10$; $2.28e-16$; (f) gEa-GAN: $8.59e-18$; $1.82e-7$; $3.72e-5$

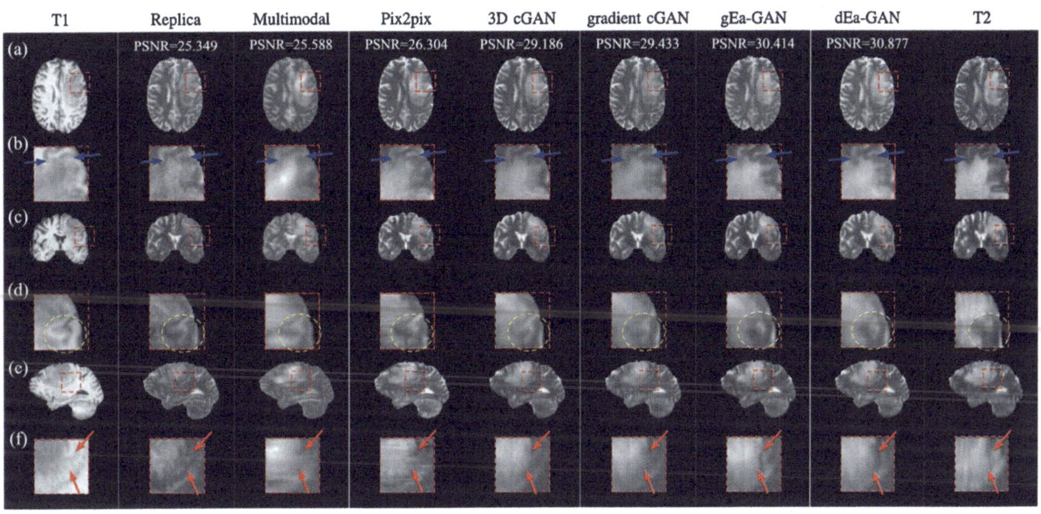

Fig. 18 Visual comparison of the synthesized T2 images between Ea-GANs and other state-of-the-art methods taken from [57]: (**a**) axial slices, (**b**) zoomed parts of axial slices, (**c**) coronal slices, (**d**) zoomed parts of coronal slices, and (**e**) sagittal slices, (**f**) zoomed parts of sagittal slices

works, we can see that using the recent GAN based models achieves better medical image synthesis performance than the conventional CNN based models. Besides, we can also conclude two main factors that benefit the successful application of the presented GAN models in within-modality and cross-modality synthesis. First, due to the 3D structure of medical images, the 3D architecture of GAN models can preserve continuous contextual information along all the three directions and therefore improve the synthesis results. Second, in order to synthesize more realistic images, additionally exploiting the spatially local details in the source or the target images for different subjects can further boost the synthesis performance of the GAN based methods, since the subtle visual difference in medical images is essential in clinical applications. In summary, deep learning based medical image synthesis, especially the recent GAN based one, has become an active research topic. With the participation of more researchers, it is expected that new synthesis approaches and methods will be developed in the coming years to further boost its performance and efficiency.

References

1. An L, Zhang P, Adeli E, Wang Y, Ma G, Shi F, Lalush DS, Lin W, Shen D (2016) Multi-level canonical correlation analysis for standard-dose PET image estimation. IEEE Trans Image Process 25(7):3303–3315
2. Armanious K, Jiang C, Fischer M, Küstner T, Nikolaou K, Gatidis S, Yang B (2018) MedGAN: medical image translation using GANs. arXiv preprint. arXiv:1806.06397
3. Ben-Cohen A, Klang E, Raskin SP, Amitai MM, Greenspan H (2017) Virtual PET images from CT data using deep convolutional networks: initial results. In: International workshop on simulation and synthesis in medical imaging. Springer, Cham, pp 49–57
4. Bi L, Kim J, Kumar A, Feng D, Fulham M (2017) Synthesis of positron emission tomography (PET) images via multi-channel generative adversarial networks (GANs). In: Molecular imaging, reconstruction and analysis of moving body organs, and stroke imaging and treatment. Springer, Cham, pp 43–51
5. Bullitt E, Zeng D, Gerig G, Aylward S, Joshi S, Smith JK, Lin W, Ewend MG (2005) Vessel tortuosity and brain tumor malignancy: a blinded study1. Acad Radiol 12(10):1232–1240
6. Bustin A, Voilliot D, Menini A, Felblinger J, de Chillou C, Burschka D, Bonnemains L, Odille F (2018) Isotropic reconstruction of MR images using 3D patch-based self-similarity learning. IEEE Trans Med Imaging 37:1932–1942
7. Chartsias A, Joyce T, Dharmakumar R, Tsaftaris SA (2017) Adversarial image synthesis for unpaired multi-modal cardiac data. In: International workshop on simulation and synthesis in medical imaging. Springer, Cham, pp 3–13
8. Chartsias A, Joyce T, Giuffrida MV, Tsaftaris SA (2018) Multimodal MR synthesis via modality-invariant latent representation. IEEE Trans Med imaging 37(3):803–814
9. Chaudhari AS, Fang Z, Kogan F, Wood J, Stevens KJ, Gibbons EK, Lee JH, Gold GE, Hargreaves BA (2018) Super-resolution musculoskeletal MRI using deep learning. Magn Reson Med 80(5):2139–2154
10. Chen H, Zhang Y, Kalra MK, Lin F, Chen Y, Liao P, Zhou J, Wang G (2017) Low-dose CT with a residual encoder-decoder convolutional neural network. IEEE Trans Med Imaging 36(12):2524–2535
11. Chen H, Zhang Y, Zhang W, Liao P, Li K, Zhou J, Wang G (2017) Low-dose CT via convolutional neural network. Biomed Opt Express 8(2): 679–694
12. Chen W (2007) Clinical applications of PET in brain tumors. J Nucl Med 48(9):1468–1481
13. Choi H, Lee DS (2018) Generation of structural MR images from amyloid PET: application to MR-less quantification. J Nucl Med 59(7):1111–1117
14. Costa P, Galdran A, Meyer MI, Niemeijer M, Abràmoff M, Mendonça AM, Campilho A (2018) End-to-end adversarial retinal image synthesis. IEEE Trans Med Imaging 37(3):781–791
15. Dadar M, Pascoal TA, Manitsirikul S, Misquitta K, Fonov VS, Tartaglia MC, Breitner J, Rosa-Neto P, Carmichael OT, Decarli C et al (2017) Validation of a regression technique for segmentation of white matter hyperintensities in Alzheimer's disease. IEEE Trans Med Imaging 36:1758–1768
16. Dar SUH, Yurt M, Karacan L, Erdem A, Erdem E, Çukur T (2018) Image synthesis in multi-contrast MRI with conditional generative adversarial networks. arXiv preprint. arXiv:1802.01221
17. Emami H, Dong M, Nejad-Davarani SP, Glide-Hurst CK (2018) Generating synthetic CTS from magnetic resonance images using generative adversarial networks. Med Phys 45(8):3627–3636
18. Frangi AF, Tsaftaris SA, Prince JL (2018) Simulation and synthesis in medical imaging. IEEE Trans Med Imaging 37(3):673
19. Goodfellow I, Pouget-Abadie J, Mirza M, Xu B, Warde-Farley D, Ozair S, Courville A, Bengio Y (2014) Generative adversarial nets. In: Advances in neural information processing systems, pp 2672–2680
20. Han X (2017) MR-based synthetic CT generation using a deep convolutional neural network method. Med Phys 44(4):1408–1419
21. Hiasa Y, Otake Y, Takao M, Matsuoka T, Takashima K, Carass A, Prince JL, Sugano N, Sato Y (2018) Cross-modality image synthesis from unpaired data using cycleGAN. In: International workshop on simulation and synthesis in medical imaging. Springer, Cham, pp 31–41
22. Hsieh J et al (2009) Computed tomography: principles, design, artifacts, and recent advances. SPIE, Bellingham, WA
23. Huang Y, Shao L, Frangi AF (2017) Cross-modality image synthesis via weakly coupled and geometry co-regularized joint dictionary learning. IEEE Trans Med Imaging 37(3):815–827
24. Huang Y, Shao L, Frangi AF (2017) Simultaneous super-resolution and cross-modality synthesis of 3D medical images using weakly-supervised joint convolutional sparse coding. arXiv preprint. arXiv:1705.02596
25. Isola P, Zhu JY, Zhou T, Efros AA (2016) Image-to-image translation with conditional adversarial networks. arXiv preprint. arXiv:1611.07004
26. Jog A, Carass A, Roy S, Pham DL, Prince JL (2017) Random forest regression for magnetic resonance image synthesis. Med Image Anal 35:475–488
27. Kamnitsas K, Ledig C, Newcombe VF, Simpson JP, Kane AD, Menon DK, Rueckert D, Glocker B (2017) Efficient multi-scale 3D CNN with fully connected

CRF for accurate brain lesion segmentation. Med Image Anal 36:61–78
28. Kang E, Min J, Ye JC (2017) A deep convolutional neural network using directional wavelets for low-dose x-ray CT reconstruction. Med Phys 44(10):e360–e375
29. Krizhevsky A, Sutskever I, Hinton GE (2012) ImageNet classification with deep convolutional neural networks. In: Advances in neural information processing systems, pp 1097–1105
30. Lê M, Delingette H, Kalpathy-Cramer J, Gerstner ER, Batchelor T, Unkelbach J, Ayache N (2017) Personalized radiotherapy planning based on a computational tumor growth model. IEEE Trans Med Imaging 36(3):815–825
31. Leynes AP, Yang J, Wiesinger F, Kaushik SS, Shanbhag DD, Seo Y, Hope TA, Larson PE (2017) Direct pseudoCT generation for pelvis PET/MRI attenuation correction using deep convolutional neural networks with multi-parametric MRI: zero echo-time and dixon deep pseudoCT (ZeDD-CT). J Nucl Med 59:852–858 (2017)
32. Liu F, Jang H, Kijowski R, Bradshaw T, McMillan AB (2017) Deep learning MR imaging–based attenuation correction for PET/MR imaging. Radiology 286(2):676–684
33. Menze BH, Jakab A, Bauer S, Kalpathy-Cramer J, Farahani K, Kirby J, Burren Y, Porz N, Slotboom J, Wiest R et al (2015) The multimodal brain tumor image segmentation benchmark (brats). IEEE Trans Med Imaging 34(10):1993–2024
34. Mirza M, Osindero S (2014) Conditional generative adversarial nets. arXiv preprint. arXiv:1411.1784
35. Mok TC, Chung AC (2018) Learning data augmentation for brain tumor segmentation with coarse-to-fine generative adversarial networks. In: International MICCAI brainlesion workshop. Springer, Cham, pp 70–80
36. Nie D, Cao X, Gao Y, Wang L, Shen D (2016) Estimating CT image from MRI data using 3D fully convolutional networks. In: Deep learning and data labeling for medical applications. Springer, Cham, pp 170–178
37. Nie D, Trullo R, Lian J, Wang L, Petitjean C, Ruan S, Wang Q, Shen D (2018) Medical image synthesis with deep convolutional adversarial networks. IEEE Trans Biomed Eng 65:2720–2730
38. Olut S, Sahin YH, Demir U, Unal G (2018) Generative adversarial training for MRA image synthesis using multi-contrast MRI. In: International workshop on predictive intelligence in medicine. Springer, Cham, pp 147–154
39. Peterfy C, Schneider E, Nevitt M (2008) The osteoarthritis initiative: report on the design rationale for the magnetic resonance imaging protocol for the knee. Osteoarthr Cartil 16(12):1433–1441
40. Ronneberger O, Fischer P, Brox T (2015) U-net: convolutional networks for biomedical image segmentation. In: International conference on medical image computing and computer-assisted intervention. Springer, Cham, pp 234–241
41. Sahiner B, Pezeshk A, Hadjiiski LM, Wang X, Drukker K, Cha KH, Summers RM, Giger ML (2019) Deep learning in medical imaging and radiation therapy. Med Phys 46(1):e1–e36
42. Van Nguyen H, Zhou K, Vemulapalli R (2015) Cross-domain synthesis of medical images using efficient location-sensitive deep network. In: International conference on medical image computing and computer-assisted intervention. Springer, Cham. pp 677–684
43. van Tulder G, de Bruijne M (2015) Why does synthesized data improve multi-sequence classification? In: International conference on medical image computing and computer-assisted intervention. Springer, Cham, pp 531–538
44. Wang Y, Zhang P, An L, Ma G, Kang J, Shi F, Wu X, Zhou J, Lalush DS, Lin W et al (2016) Predicting standard-dose PET image from low-dose PET and multimodal MR images using mapping-based sparse representation. Phys Med Biol 61(2):791
45. Wang Y, Ma G, An L, Shi F, Zhang P, Lalush DS, Wu X, Pu Y, Zhou J, Shen D (2016) Semisupervised tripled dictionary learning for standard-dose PET image prediction using low-dose PET and multimodal MRI. IEEE Trans Biomed Eng 64(3): 569–579
46. Wang Y, Yu B, Wang L, Zu C, Lalush DS, Lin W, Wu X, Zhou J, Shen D, Zhou L (2018) 3D conditional generative adversarial networks for high-quality PET image estimation at low dose. NeuroImage 174:550–562
47. Wang Y, Zhou L, Yu B, Wang L, Zu C, Lalush DS, Lin W, Wu X, Zhou J, Shen D (2019) 3D auto-context-based locality adaptive multi-modality GANs for PET synthesis. IEEE Trans Med Imaging 38(6):1328
48. Wang Z, Bovik AC, Sheikh HR, Simoncelli EP (2004) Image quality assessment: from error visibility to structural similarity. IEEE Trans Image Process 13(4):600–612
49. Wei W, Poirion E, Bodini B, Durrleman S, Ayache N, Stankoff B, Colliot O (2018) Learning myelin content in multiple sclerosis from multimodal MRI through adversarial training. In: International conference on medical image computing and computer-assisted intervention. Springer, Cham, pp 514–522
50. Welander P, Karlsson S, Eklund A (2018) Generative adversarial networks for image-to-image translation on multi-contrast MR images-a comparison of cycleGAN and unit. arXiv preprint. arXiv:1806.07777
51. Xiang L, Qiao Y, Nie D, An L, Lin W, Wang Q, Shen D (2017) Deep auto-context convolutional neural networks for standard-dose PET image estimation from low-dose PET/MRI. Neurocomputing 267:406–416
52. Yang Q, Li N, Zhao Z, Fan X, Chang EI, Xu Y et al (2018) MRI cross-modality neuroimage-to-neuroimage translation. arXiv preprint. arXiv:1801.06940

53. Ye DH, Zikic D, Glocker B, Criminisi A, Konukoglu E (2013) Modality propagation: coherent synthesis of subject-specific scans with data-driven regularization. In: International conference on medical image computing and computer-assisted intervention. Springer, Cham, pp 606–613
54. Yi X, Babyn P (2017) Sharpness-aware low dose CT denoising using conditional generative adversarial network. arXiv preprint. arXiv:1708.06453
55. Yi X, Walia E, Babyn P (2018) Generative adversarial network in medical imaging: a review. arXiv preprint. arXiv:1809.07294
56. Yu B, Zhou L, Wang L, Fripp J, Bourgeat P (2018) 3D cGAN based cross-modality MR image synthesis for brain tumor segmentation. In: 2018 IEEE 15th international symposium on biomedical imaging (ISBI 2018). IEEE, Washington, DC, pp 626–630
57. Yu B, Zhou L, Wang L, Shi Y, Fripp J, Bourgeat P (2019) Ea-GANs: edge-aware generative adversarial networks for cross-modality MR image synthesis. IEEE Trans Med Imaging 38:1750–1762
58. Zeng K, Zheng H, Cai C, Yang Y, Zhang K, Chen Z (2018) Simultaneous single-and multi-contrast super-resolution for brain MRI images based on a convolutional neural network. Comput Biol Med 99:133–141
59. Zhang Z, Yang L, Zheng Y (2018) Translating and segmenting multimodal medical volumes with cycle-and shape-consistency generative adversarial network. In: Proceedings of the IEEE conference on computer vision and pattern recognition, pp 9242–9251

Part II

Applications: Screening and Diagnosis

Deep Learning for Pulmonary Image Analysis: Classification, Detection, and Segmentation

Shoji Kido, Yasushi Hirano, and Shingo Mabu

Abstract

Image-based computer-aided diagnosis (CAD) algorithms by the use of convolutional neural network (CNN) which do not require the image-feature extractor are powerful compared with conventional feature-based CAD algorithms which require the image-feature extractor for classification of lung abnormalities. Moreover, computer-aided detection and segmentation algorithms by the use of CNN are useful for analysis of lung abnormalities. Deep learning will improve the performance of CAD systems dramatically. Therefore, they will change the roles of radiologists in the near future. In this article, we introduce development and evaluation of such image-based CAD algorithms for various kinds of lung abnormalities such as lung nodules and diffuse lung diseases.

S. Kido (✉)
Graduate School of Medicine, Osaka University, Suita, Osaka, Japan
e-mail: kido@radiol.med.osaka-u.ac.jp

Y. Hirano · S. Mabu
Graduate School of Sciences and Technology for Innovation, Yamaguchi University, Ube, Yamaguchi, Japan
e-mail: yhirano@yamaguchi-u.ac.jp; mabu@yamaguchi-u.ac.jp

Keywords

Computer-aided diagnosis (CAD) · Convolutional neural network (CNN) · R-CNN · Fully convolutional network (FCN) · U-Net · Residual U-Net · V-Net · Lung nodule · Diffuse lung disease

Background of Lung Diseases

Lung diseases include many disorders affecting lungs, such as infections like pneumonia, chronic obstructive lung disease (COPD), lung cancer, and interstitial lung diseases. There are various factors that cause lung diseases. Viruses and bacteria cause pneumonia. Mutations cause lung cancer. Especially smoking causes various lung diseases such as lung cancer, COPD, and so on. Many people in the world are killed by lung diseases. In the top ten global causes of death, 2016, the third was COPD, fourth was infections, sixth was lung cancer, and tenth was tuberculosis [1]. In general, early diagnosis of lung diseases improves the prognosis. For example, in the national lung screening trial of USA, CT screening was able to reduce lung cancer deaths by 20% [2].

Lung diseases in images are roughly divided into two categories, one is localized diseases such

G. Lee, H. Fujita (eds.), *Deep Learning in Medical Image Analysis*, Advances in Experimental Medicine and Biology 1213, https://doi.org/10.1007/978-3-030-33128-3_3

as lung nodules, and the other is diffuse lung diseases. A lung nodule is a localized lesion surrounded by lung. Lung nodules include not only malignant nodules like lung cancer but also benign nodules. By using X-ray images, it is difficult for radiologists to detect small nodules or nodules overlapping ribs or mediastinum. By using CT images, such nodules are easy to detect compared with X-ray images. However, many small nodules can be detected in CT lung cancer screening, and this burdens to radiologists. So, the CAD system to diagnose lung nodules is expected for radiologists. The lung nodules are divided into three types based on the proportion that contains solid component and ground-glass opacity (GGO) component: solid nodule, sub-solid nodule, and GGO nodule. Sub-solid nodule includes both of solid and GGO components. The size of the nodule and the proportion of GGO component are important for diagnosing benign or malignant nodules. Diffuse lung diseases spread both the left and right lungs and contain many diseases such as interstitial lung diseases, infectious diseases, and lung cancers. They reveal a variety of opacity patterns such as consolidation, GGO, honeycombing, emphysema, and nodular. These opacity patterns are important for the diagnosis of diffuse lung diseases. So, the CAD system to diagnose lung nodules is also expected for radiologists.

Introduction

The image recognition ability of the deep learning approach was noticed in the 2012 ImageNet Large-Scale Visual Recognition Challenge (ILSVRC), in which convolutional neural networks (CNNs), one of the deep learning models, were first used. The image recognition accuracy of Hinton's team from the University of Toronto gained substantial attention as a result of beating the other teams by 10% or more [3]. All top winners in the ILSVRC events since then have been based on CNNs. Furthermore, in the 2015 ILSVRC, the team of Microsoft Research Asia achieved a misclassification rate of 3.6% using CNN of 152 layers, which was 5.1% lower than human misclassification rate. The recognition ability of CNN is already considered to be superior to human recognition ability in natural image classification, and high discrimination ability is expected in recognition in medical image diagnosis. We have developed and evaluated the performance of such image-based CAD algorithms by the use of CNN for various kinds of lung abnormalities such as lung nodules and diffuse lung diseases.

For assisting radiologists' diagnoses, CAD systems include three types of algorithms such as "classification" that differentiate abnormal lesions into benign or malignant, or histological subtypes, "detection" that find abnormal lesions, and "segmentation" that extract organs or abnormal areas (Fig. 1). In usual CAD algorithms, designing an image-feature extractor is important. However, this task is difficult for medical engineers. On the other hand, CAD algorithm by the use of CNN does not require the image-feature extractor (Fig. 2). In this study, we have developed image-based CAD algorithms for classification, detection, and segmentation of lung abnormalities by the use of CNN. We evaluated the performance of such CAD algorithms for various kinds of lung abnormalities such as lung nodules and diffuse lung diseases.

Methods

Classification of Lung Abnormalities

Classification is to classify the main object category such as a lung nodule or diffuse lung opacity within an image. When classification of diffuse lung opacities is performed by CNN, it is not necessary to design image features for each opacity. We classified diffuse lung opacities into five representative abnormal opacity patterns (consolidation, ground-glass opacity (GGO), honeycombing, emphysema, diffuse nodular) and normal lung using high-resolution CT (HRCT) images. As training data for CNN, we selected slices of HRCT images with each of diffuse lung opacity and three radiologists independently annotated areas where each opacity was considered to be present. We extracted regions where at least two

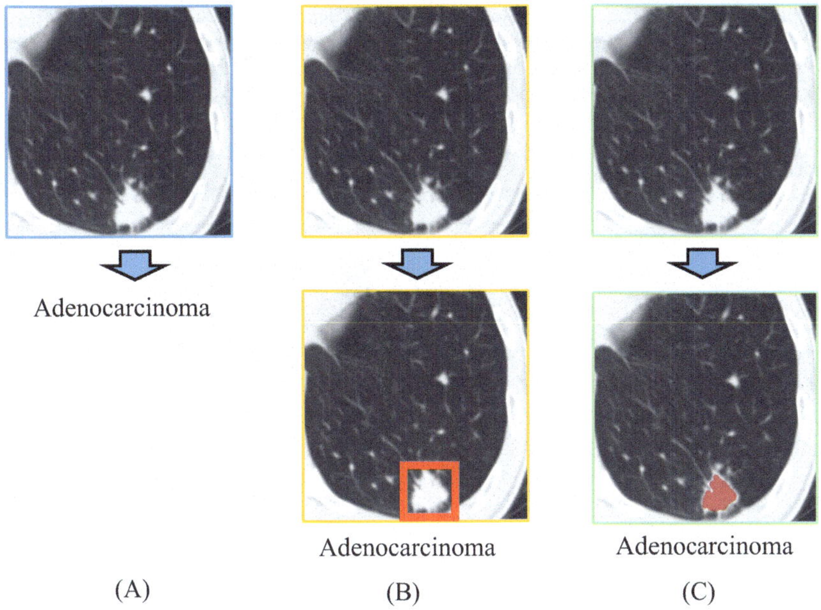

Fig. 1 Three types of CAD algorithms for assisting radiologists' diagnoses. (**a**) Classification, (**b**) Detection, (**c**) Segmentation

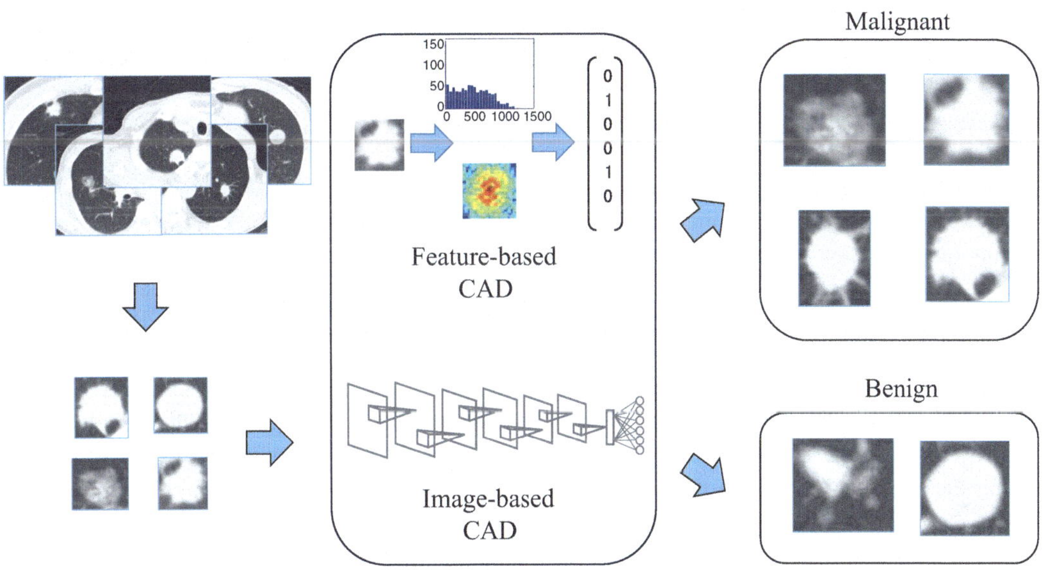

Fig. 2 For classification of lung abnormalities, conventional feature-based CAD algorithm requires the image-feature extractor, however, image-based CAD algorithm by the use of CNN does not require the image-feature extractor

of three radiologists' annotations coincided as the ground truth (Fig. 3). For the ground truth area, we set the region of interest (ROI) of 32 pixels on one side for each opacity pattern. The number of ROIs obtained at this point was different for each opacity, so it was unified to 500. However, since the number of ROIs of about 500 is not enough to train CNN, data augmentation of the training data by rotation and reflection is performed to prepare 4000 training data for each opacity. The CNN

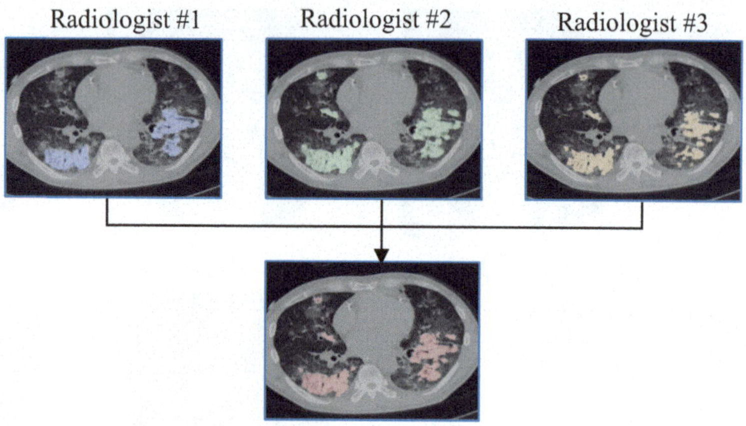

Fig. 3 Annotation of diffuse lung opacities. At first, three radiologists annotated each slice independently. Ground truth was defined as a coincided annotated area by two or more of three radiologists

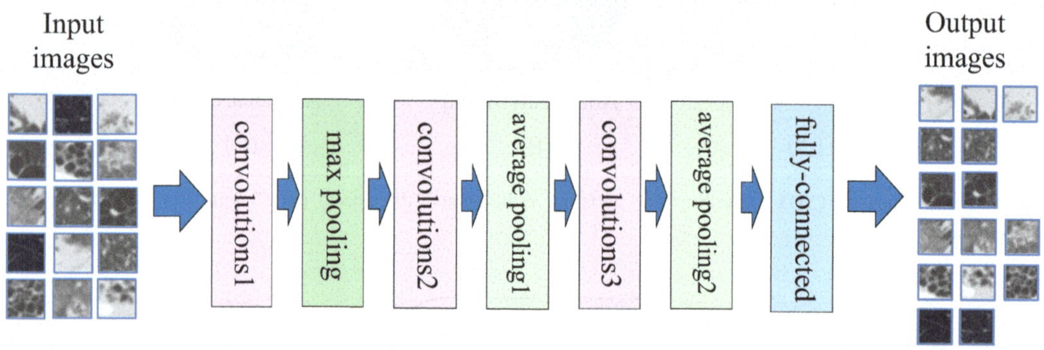

Fig. 4 Proposed CNN model. The CNN model we used had three convolution layers, max-pooling layer, two average pooling layers, and a fully connected layer

model we used had three convolution layers, max-pooling layer, two average pooling layers, and a fully connected layer (Fig. 4). As a result of the fivefold cross-validation, the mean discrimination rate of 84.7 ± 0.7% was obtained (Table 1). In addition, since the distribution of opacities in the lungs is also important in the diagnosis of diffuse lung diseases, we classified opacity patterns by sliding the ROI to all pixels in the lung. It was shown that this result coincided well with the radiologists' annotation (Fig. 5) [4].

The benign/malignant classification of lung nodules is a common and important task in daily clinical practice. Figure 6 shows benign and malignant lung nodules on HRCT images used for comparison of a feature-based CAD using speeded up robust features (SURF) and bag-of-features, and an image-based CAD using CNN [5]. In this study, when classification was performed using feature-based CAD, the classification rate was almost the same as the random selection of 55%. However, image-based CAD achieved a high classification rate of 86% (Table 2). Furthermore, when benign and malignant nodules were divided into solid and sub-solid types, respectively, the classification rate of feature-based CAD was the same as the random selection at 25%. While image-based CAD was 58% compared to the conventional method (Table 3). Since the number of nodules used in this experiment was relatively small, transfer learning with AlexNet [3] was used. Transfer learning is a method that uses a pre-trained model with large-scale image data such as ImageNet, and is a useful method when collecting large-scale images is difficult, such as medical images. The result indicated the usefulness of image-based CAD for classification of benign and malignant lung nodules. Since lung nodules have a three-dimensional (3D) structure, CNN-based image classification using 3D input images is considered to be useful. We

Table 1 The classification result of diffuse lung opacities using CNN

Estimation by CNN (columns); Diagnosis by radiologists (rows)

	CON	GGO	HCM	EMP	NOD	NOR	Accuracy (%)
CON	**3828**	86	73	2	0	11	**95.7±0.4**
GGO	102	**3340**	183	32	254	89	**83.5±1.1**
HCM	81	160	**3593**	152	14	0	**89.8±0.7**
EMP	7	37	221	**3431**	201	103	**85.8±0.9**
NOD	15	212	36	103	**2929**	705	**73.2±0.9**
NOR	2	68	11	78	630	**3211**	**80.3±1.0**

Mean accuracy: 84.7±0.7%

CON consolidation, *GGO* ground-glass opacity, *HCM* honeycombing, *EMP* emphysema, *NOD* diffuse nodular, *NOR* normal

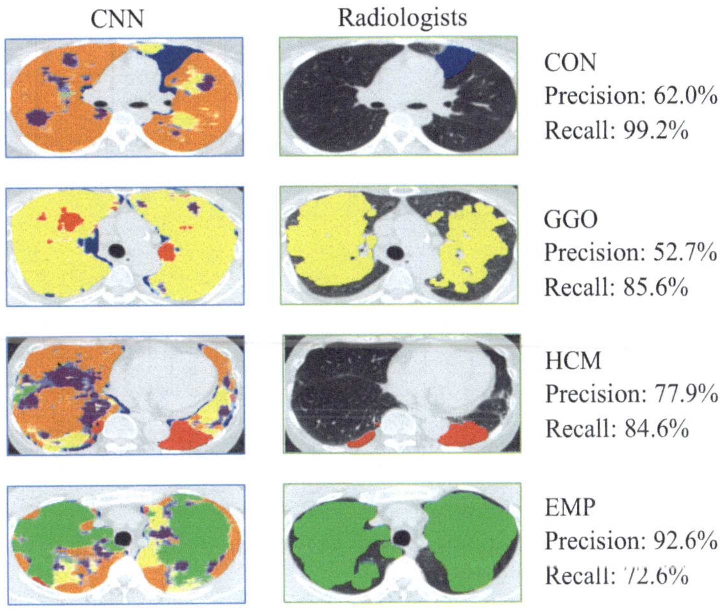

Fig. 5 Classification of diffuse lung opacities by sliding the ROI to all pixels in the lung. The classification results indicate each diffuse lung opacity patterns resemble radiologists' annotations. In each image, radiologists annotated only one of the diffuse lung opacities. *CON* consolidation, *GGO* ground-glass opacity, *HCM* honeycombing, *EMP* emphysema

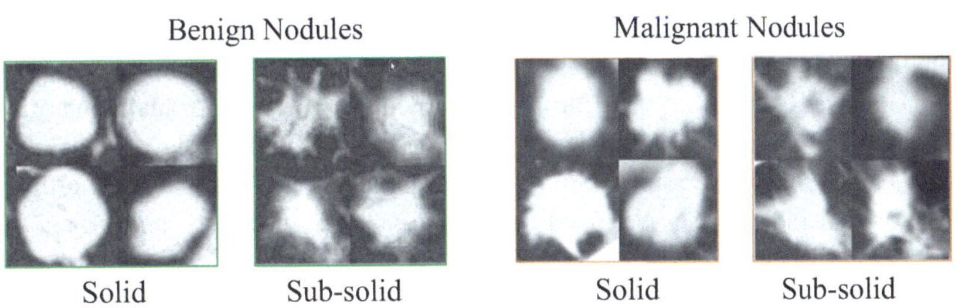

Fig. 6 Benign and malignant lung nodules on HRCT images. Nodules can be divided into solid and sub-solid types, respectively

Table 2 Comparison of feature-based CAD and image-based CAD

Feature-based CAD	Benign	Malignant
Benign	**13**	8
Malignant	10	**10**

Image-based CAD	Benign	Malignant
Benign	**19**	2
Malignant	4	**17**

Accuracy = 55% < Accuracy = 86%

Classification of benign and malignant nodules

Table 3 Comparison of feature-based CAD and image-based CAD

Feature-based CAD	B-Solid	B-Sub	M-Solid	M-Sub
B-Solid	**2**	2	2	0
B-Sub	2	**2**	0	2
M-Solid	2	1	**1**	2
M-Sub	1	3	1	**1**

Image-based CAD	B-Solid	B-Sub	M-Solid	M-Sub
B-Solid	**4**	2	0	0
B-Sub	1	**2**	1	2
M-Solid	0	0	**5**	1
M-Sub	0	0	3	**3**

Accuracy = 25% < Accuracy = 58%

Classification of solid and sub-solid of lung nodules. *B-Solid* benign solid, *B-Sub* benign sub-solid, *M-Solid* malignant solid, *M-Sub* malignant sub-solid

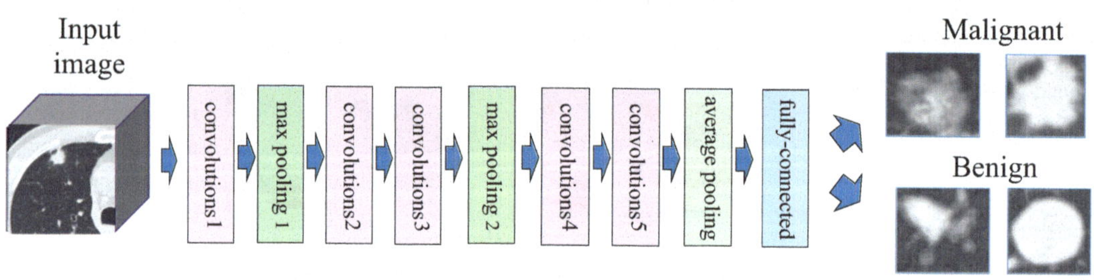

Fig. 7 Proposed CNN model. The CNN model we used had five convolution layers, two max-pooling layers, an average pooling layer, and a fully connected layer

attempted to discriminate between benign and malignant lung nodules using the images of The Lung Image Database Consortium and Image Database Resource Initiative (LIDC-IDRI). The lung nodules we used were 635 rated in five grades by three or four radiologists. We used the 3D CNN model with five convolution layers, two max-pooling layers, an average pooling layer, and a fully connected layer to classify benign or malignant nodules (Fig. 7). The result of threefold cross-validation was 79.4 ± 3.5%, and the results were well correlated with the radiologists' evaluation [6]. Analysis using 3D images is useful to evaluate the whole nodule, but using 3D images as input has a large computational cost with a limited GPU memory, and also it cannot use pre-trained networks for transfer learning. For this reason, some methods have been devised that use the largest split planes from multiple directions such as axial, coronal, sagittal, etc. as the input images [7].

Detection of Lung Abnormalities

Detection is to classify the object category and locate the position using a bounding box for every object within an image. In 2014, Girshick et al.

proposed a regions with CNN features (R-CNN) to which the CNN algorithm was applied [8]. The following year, Fast R-CNN [9] and Faster R-CNN [10], an improvement of R-CNN, were proposed, and You Only Look Once (YOLO) [11] and Single Shot MultiBox Detector (SSD) [12] were also proposed.

R-CNN is an object detection framework, which uses a CNN to classify regions within an image. In our previous study, we used a sliding window for the classification of lung abnormalities on CT images. Instead of using a sliding window, the R-CNN detector only deals with those regions that are likely to contain an object. In R-CNN, a large number of object candidate areas are set based on gray scale information or texture information in the images, these object candidate areas are resized to a fixed size, and the CNN performs feature extraction and the object position in the image (Fig. 8). The results of applying this method to lung nodules (Fig. 9) and diffuse lung diseases (Fig. 10) are shown [13]. In the conventional method, when the nodule attached to the chest wall or stuck in the azygoesophageal recess, it is necessary to devise an algorithm for separating the nodule from the chest wall attached to it. In addition, it is necessary to devise an algorithm for detecting nodules with air-bronchogram or GGO. However, such a device is unnecessary in the method using R-CNN. In addition, even in the case of diffuse lung diseases, it is possible to detect opacities as shown in the figure without defining the features for each opacity pattern.

Segmentation of Lung Abnormalities

Segmentation is to identify the object category of each pixel for every known object within an image. As an algorithm for image segmentation, fully convolutional network (FCN), which is

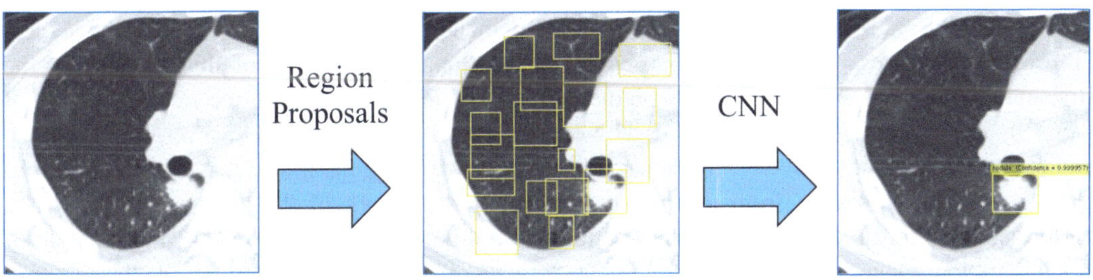

Fig. 8 Lung nodule detection by R-CNN. In the first step, candidate regions are generated (region proposals), and in the second step, lung nodule is determined by the use of CNN

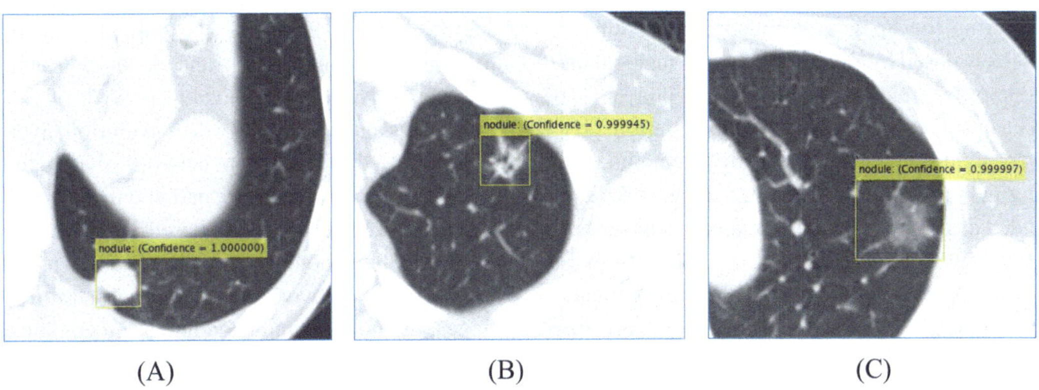

Fig. 9 Examples of lung nodules detected by R-CNN. (**a**) Nodule attached to the chest wall, (**b**) Nodule with air-bronchogram, (**c**) GGO nodule

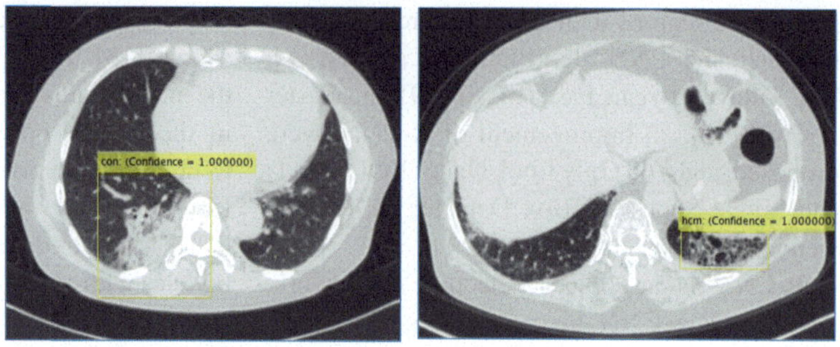

(A) (B)

Fig. 10 Examples of diffuse lung opacities detected by R-CNN. (**a**) Consolidation, (**b**) Honeycombing

a CNN not using a fully connected layer, was proposed by Long et al. [14], and the FCN architecture became widespread. Region extraction can be performed with images of any size, and many modified architectures have been announced since then.

For diffuse lung opacities, it is possible to segment diffuse lung opacities by sliding the ROI to all pixels and to classify the opacity patterns. However, with this method, we need to set the fixed size of the ROI, which cannot be changed freely for each pixel. On the other hand, we used U-Net to segment each opacity pattern of diffuse lung diseases. U-Net was proposed by Ronneberger et al. for the purpose of segmentation in medical images in 2015 [15]. U-Net has an encoder-decoder structure. In the encoder part, the feature extraction from the images is performed, and in the decoder part, the feature is held and the image is restored. Since the corresponding feature maps of the encoder part and the decoder part are connected, there is a feature that position information can be included when restoring the images. We also used residual U-Net, proposed by Zhang et al. using residual units to improve training problems in deep layers [16]. The opacities we used were consolidation, GGO, honeycombing, emphysema, diffuse nodular, and normal lungs. As the training data, we used coincided areas annotated by three radiologists as we used in the classification by the use of CNN [4]. Furthermore, data augmentation using reflection and homography transformation was performed on these images, and the same number of images were randomly selected for each opacity as training data. The structure of U-Net proposed by us has 19 convolution layers and 4 max-pooling layers (Fig. 11). The structure of residual U-Net proposed by us has 11 convolution layers and the two max-pooling layers were configured to be shallower than U-Net (Fig. 12). The results were in good agreement with the radiologist's annotation (Fig. 13). The average of Dice coefficients for all opacities were 0.800 ± 0.101 for U-Net, and 0.854 ± 0.065 for residual U-Net (Table 4) [17].

Noh et al. proposed deconvolution network [18] in 2015, and Milletari et al. proposed V-Net [19] in 2016, both models are for image segmentation. Like the CNN used in image classification, the first half of this network is the convolution network, and the image features of the target are extracted. However, the area map is created using the image features in the deconvolution network of the second half (Fig. 14). V-Net is a method proposed for the purpose of medical image segmentation as well as U-Net. While U-Net targets two-dimensional images, V-Net targets 3D images (Fig. 15). As the difference between the two models, deconvolution network uses the pooling layer and the unpooling layer during image compression and decompression. However, V-Net uses a convolution layer of two kernel sizes and strides and a deconvolution layer. In addition, deconvolution network shares the pooling index in the encoder part and the decoder part, and the feature map is shared in V-Net. We segmented lung nodules from 3D images using deconvolution network and V-Net. As training images, the images annotated under the guidance of a radiologist were used, which were cut out into $128 \times 128 \times 64$ pixels centering on the gravity of the nodule. The extraction results of

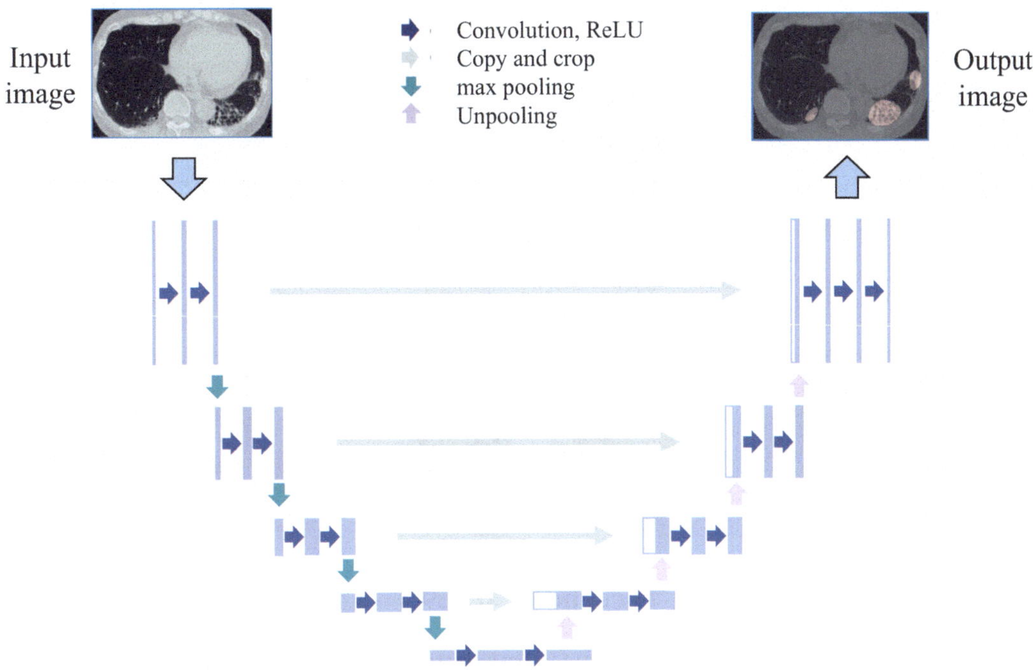

Fig. 11 Proposed U-Net model. The U-Net model we used had 19 convolution layers and 4 max-pooling layers

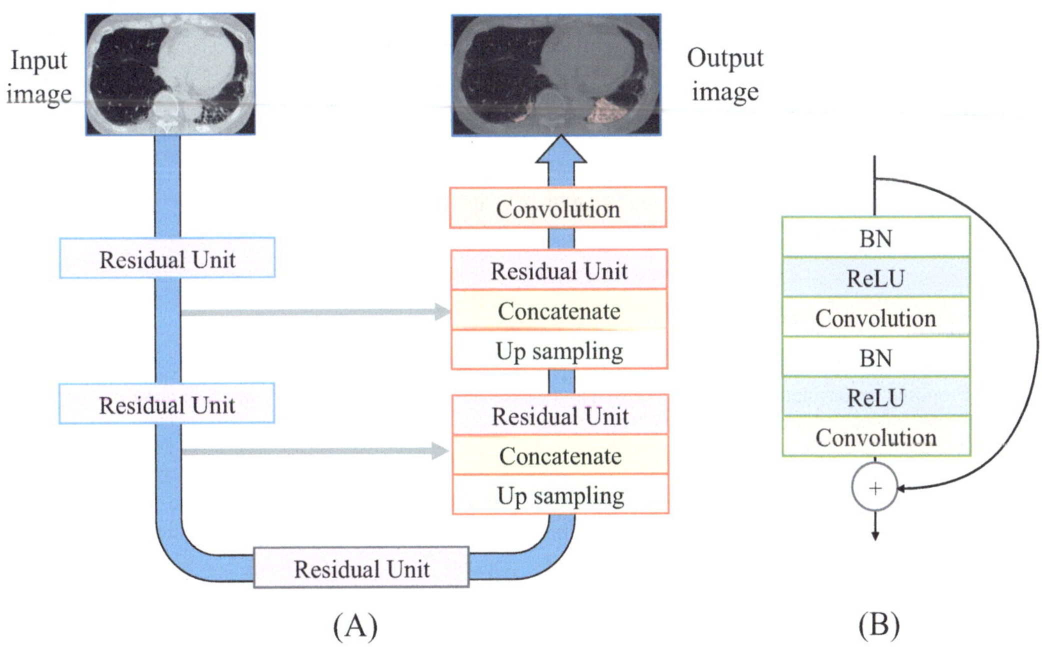

Fig. 12 Proposed residual U-Net model. (**a**) The residual U-Net model we used had 11 convolution layers and two max-pooling layers. (**b**) The structure of a residual unit. *BN* Batch normalization

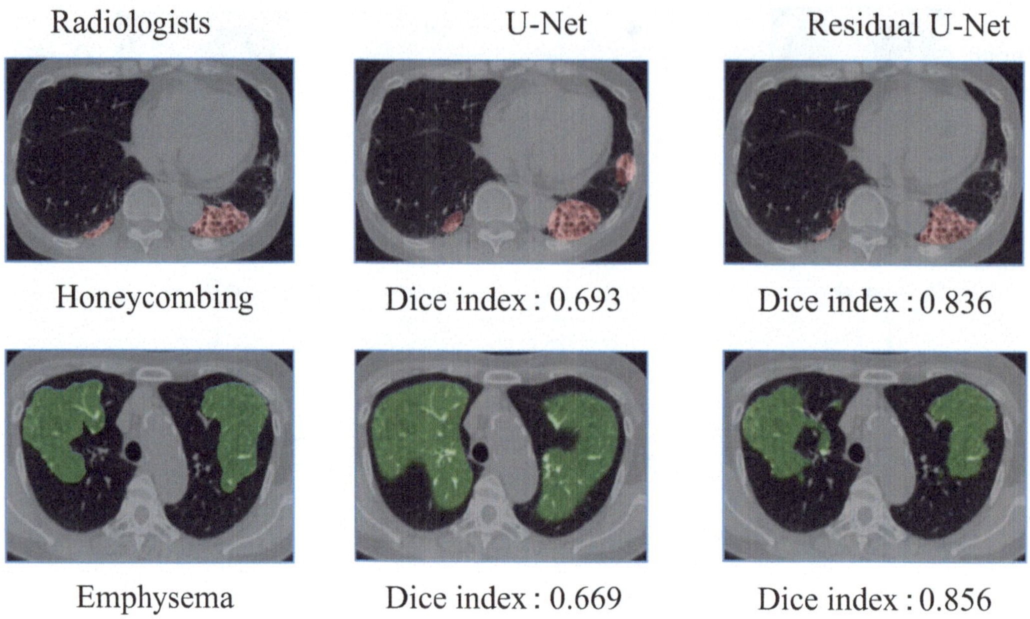

Fig. 13 Examples of segmentation results by the use of U-Net and residual U-Net. Segmentation results coincided well with the radiologists' annotation

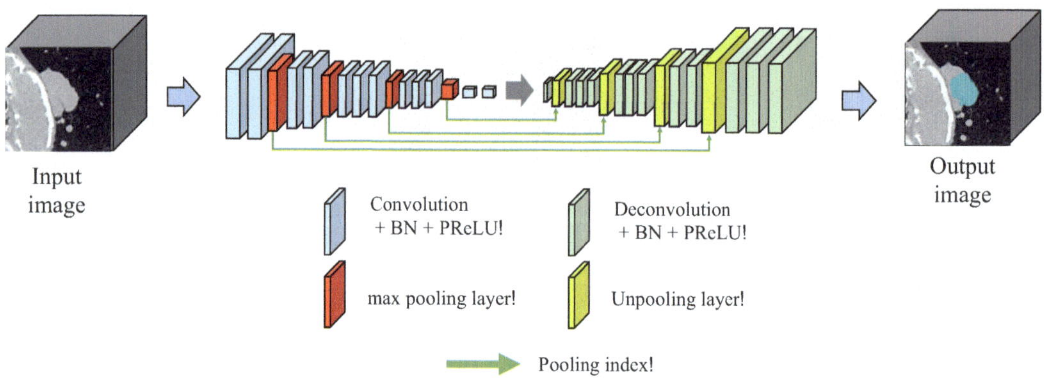

Fig. 14 Proposed deconvolution network model. The first half of this network is the convolution network, and the second half is the deconvolution network. *BN* Batch normalization

deconvolution network and V-Net coincided well with the radiologist's annotation (Fig. 16). The average Dice coefficients were 0.740 ± 0.012 and 0.810 ± 0.016, respectively [20, 21].

Conclusion

In image-based CADs by the use of CNN, classification of lung abnormalities was superior to feature-based CAD. Moreover, image-based

Table 4 Comparison of segmentation accuracy for diffuse lung opacities by use of U-Net and residual U-Net

Opacity	Dice index	
	U-Net	Residual U-Net
CON	0.911 ± 0.024	0.854 ± 0.074
GGO	0.735 ± 0.088	0.767 ± 0.070
HCM	0.768 ± 0.090	0.871 ± 0.045
EMP	0.679 ± 0.183	0.831 ± 0.109
NOD	0.745 ± 0.139	0.820 ± 0.131
NOR	0.963 ± 0.005	0.979 ± 0.004
Mean ± SD	0.800 ± 0.101	0.854 ± 0.065

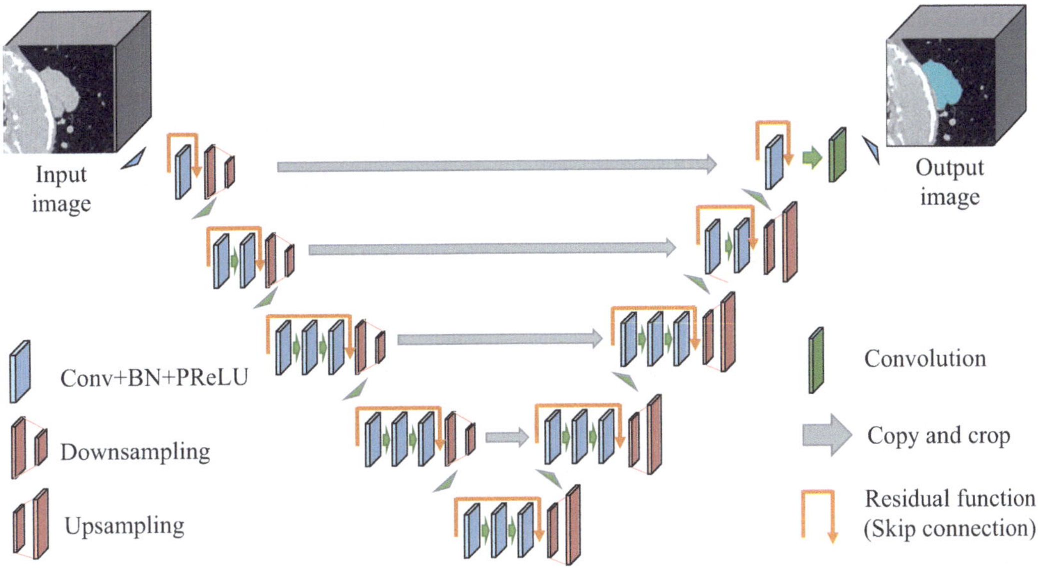

Fig. 15 Proposed V-Net model. The first half of this network is the encoder, and the second half is the decoder. *BN* Batch normalization

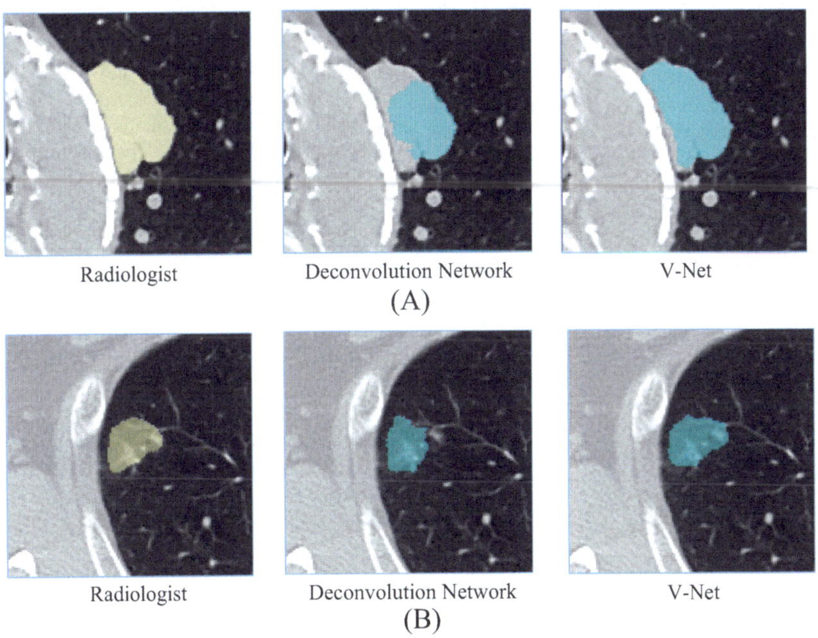

Fig. 16 Examples of segmentation results by the use of deconvolution network and V-Net. Segmentation results were coincided well with the radiologist's annotation. (**a**) A nodule attached to the aortic arch, (**b**) A GGO nodule

CAD by the use of R-CNN showed high performance for detection of lung abnormalities, and also, image-based CADs by the use of FCN, U-Net, deconvolution network, and V-Net showed high performance for segmentation of lung abnormalities. Image-based CAD algorithms are promising for classification, detection, and segmentation of various kinds of lung abnormalities such as lung nodules and diffuse lung diseases.

References

1. Global Health Estimate (2018) 2016: Death by cause, age sex, by country and by region, 2006–2016. World Health Organization, Geneva
2. The National Lung Screening Trial Research Team (2011) Reduced lung-cancer mortality with low-dose computed tomographic screening. N Eng J Med 365(5):395–409
3. Krizhevsky A, Sutskever I, Hinton GE (2012) ImageNet classification with deep convolutional neural networks. In: NIPS'12 proceedings of the 25th international conference on neural information processing systems, vol 1. Curran Associates Inc, Lake Tahoe, pp 1097–1105
4. Murakami K, Kido S, Hashimoto N et al (2017) Computer-aided classification of diffuse lung disease patterns using convolutional neural network. In: Computer assisted radiology and surgery, 31th international congress and exhibition (CARS2017). Int J Comput Assist Radiol Surg, Barcelona
5. Kido S, Hirano Y, Hashimoto N (2017) Computer-aided classification of pulmonary diseases: feature extraction based method versus non-feature extraction based method. In: Proceedings of the IWAIT2017. Institute of Electrical and Electronics Engineers (IEEE), Penang, pp 1–3
6. Ito T, Hirano Y, Kido S et al (2017) First-reader computerized system for distinction between malignant and benign nodules on thoracic CT images by means of end to end deep learning: convolutional neural network (CNN) and neural network convolution (NNC) approaches. In: RSNA 2017. Radiological Society of North America, Chicago
7. Setio AAA, Ciompi F, Litjens G et al (2016) Pulmonary nodule detection in CT images: false positive reduction using multi-view convolutional networks. IEEE Trans Med Imaging 35:1160–1169
8. Girshick R, Donahue J, Darrell T et al (2014) Rich feature hierarchies for accurate object detection and semantic segmentation. In: CVPR'14 proceedings of the 2014 IEEE conference on computer vision and pattern recognition. Institute of Electrical and Electronics Engineers (IEEE), Columbus, pp 580–587
9. Girshick R (2015) Fast R-CNN. In: The IEEE international conference on computer vision (ICCV). Institute of Electrical and Electronics Engineers (IEEE), Santiago, pp 1440–1448
10. Ren S, He K, Girshick R et al (2015) Faster R-CNN: towards real-time object detection with region proposal networks. In: NIPS'15 Proceedings of the 28th international conference on neural information processing systems, vol 1. MIT Press, Cambridge, pp 91–99
11. Redmon J, Divvala S, Girshick R et al (2016) You only look once: unified, real-time object detection. In: Proceedings of the 2016 IEEE conference on computer vision and pattern recognition (CVPR), vol 1. Institute of Electrical and Electronics Engineers (IEEE), Las Vegas, pp 779–788
12. Liu W, Anguelov D, Erhan D et al (2016) SSD: Single shot multibox detector. In: European conference on computer vision ECCV 2016: computer vision - ECCV 2016. Springer, Amsterdam, pp 21–37
13. Kido S, Hirano Y, Hashimoto N (2018) Detection and classification of lung abnormalities by the use of Convolutional Neural Network (CNN) and regions with CNN features. In: International workshop on advanced image technology 2018 (IWAIT 2018). Institute of Electrical and Electronics Engineers (IEEE), Chiang Mai. Special Session 2
14. Long J, Shelhamer E, Darrell T (2015) Fully convolutional networks for semantic segmentation. In: 2015 IEEE conference on computer vision and pattern recognition (CVPR). Institute of Electrical and Electronics Engineers (IEEE), Boston
15. Ronneberger O, Fischer P, Brox T (2015) U-Net: convolutional networks for biomedical image segmentation. In: MICCAI 2015: medical image computing and computer-assisted intervention – MICCAI 2015. Springer, Munich, pp 234–241
16. Zhang Z, Liu Q et al (2017) Road extraction by deep residual U-Net. IEEE Geosc Remote Sensing Lett 15(5):749–753. 1–5
17. Murakami K, Kido S, Hirano Y et al (2019) Segmentation for diffuse lung disease opacities on CT images using U-Net and residual U-Net. In: IEICE technical report 2019, vol 118. The Institute of Electronics, Information and Communication Engineers (IEICE), Naha, pp 175–179
18. Noh H, Hong S, Han B (2015) Learning of deconvolution network for semantic segmentation. In: ICCV'15 proceedings of the 2015 IEEE international conference on computer vision (ICCV). IEEE Computer Society, Washington, pp 1520–1528
19. Milletari F, Navab N, Ahmadi SA (2016) V-Net: Fully convolutional neural networks for volumetric medical image segmentation. In: 2016 fourth international conference on 3D vision (3DV). Institute of Electrical and Electronics Engineers (IEEE), California, pp 565–571
20. Kidera S, Kido S, Hirano Y, Mabu S, Tanaka N (2018) Segmentation of lung nodules on MDCT images by using 3D Conv-DeconvNet. In: Computer assisted radiology and surgery, 32th international congress and exhibition (CARS2018). Int J Comput Assist Radiol Surg, Barcelona
21. Kidera S, Kido S, Hirano Y, Tanaka N (2019) Segmentation of lung nodules on 3D CT images by using DeconvNet and V-Net. In: IEICE technical report 2019, vol 118. The Institute of Electronics, Information and Communication Engineers (IEICE), Naha, pp 103–106

Deep Learning Computer-Aided Diagnosis for Breast Lesion in Digital Mammogram

Mugahed A. Al-antari, Mohammed A. Al-masni, and Tae-Seong Kim

Abstract

For computer-aided diagnosis (CAD), detection, segmentation, and classification from medical imagery are three key components to efficiently assist physicians for accurate diagnosis. In this chapter, a completely integrated CAD system based on deep learning is presented to diagnose breast lesions from digital X-ray mammograms involving detection, segmentation, and classification. To automatically detect breast lesions from mammograms, a regional deep learning approach called You-Only-Look-Once (YOLO) is used. To segment breast lesions, full resolution convolutional network (FrCN), a novel segmentation model of deep network, is implemented and used. Finally, three conventional deep learning models including regular feedforward CNN, ResNet-50, and InceptionResNet-V2 are separately adopted and used to classify or recognize the detected and segmented breast lesion as either benign or malignant. To evaluate the integrated CAD system for detection, segmentation, and classification, the publicly available and annotated INbreast database is used over fivefold cross-validation tests. The evaluation results of the YOLO-based detection achieved detection accuracy of 97.27%, Matthews's correlation coefficient (MCC) of 93.93%, and F1-score of 98.02%. Moreover, the results of the breast lesion segmentation via FrCN achieved an overall accuracy of 92.97%, MCC of 85.93%, Dice (F1-score) of 92.69%, and Jaccard similarity coefficient of 86.37%. The detected and segmented breast lesions are classified via CNN, ResNet-50, and InceptionResNet-V2 achieving an average overall accuracies of 88.74%, 92.56%, and 95.32%, respectively. The performance evaluation results through all stages of detection, segmentation, and classification show that the integrated CAD system outperforms the latest conventional deep learning methodologies. We conclude that our CAD system could be used to assist

Mugahed A. Al-antari and Mohammed A. Al-masni contributed equally with all other contributors.

M. A. Al-antari
Department of Biomedical Engineering, College of Electronics and Information, Kyung Hee University, Yongin, Republic of Korea

Department of Biomedical Engineering, Sana'a Community College, Sana'a, Republic of Yemen

M. A. Al-masni · T.-S. Kim (✉)
Department of Biomedical Engineering, College of Electronics and Information, Kyung Hee University, Yongin, Republic of Korea
e-mail: tskim@khu.ac.kr

G. Lee, H. Fujita (eds.), *Deep Learning in Medical Image Analysis*, Advances in Experimental Medicine and Biology 1213, https://doi.org/10.1007/978-3-030-33128-3_4

radiologists over all stages of detection, segmentation, and classification for diagnosis of breast lesions.

Keywords
Medical image analysis · Mammograms · Breast lesion · Computer-aided diagnosis (CAD) · Deep learning · Full resolution convolutional network (FrCN) · Detection · Segmentation · Classification

Introduction

Breast cancer is the second most common cancer affecting the life of women worldwide [1]. In 2019, breast cancer is statistically categorized among the highest causes of all other cancers, accounting for 30% of estimated new cases and 15% of death cases [1]. Early detection of breast cancer is a crucial requirement to reduce the mortality rate among women [1]. Up to date, digital X-ray mammography is the standard and most reliable tool to screen out the suspicious breast masses and microcalcifications for patients [2]. In 2015, the American Cancer Society (ACS) updated its breast cancer screening roles. Currently, ACS recommends women over 45 years to undergo breast screening one time per year using double views of mammograms: Mediolateral Oblique (MLO) and Cranio-Caudal (CC). Whereas, women with age of 54 years or elder are encouraged to undergo breast screening every 2 years [3]. In the diagnosis of breast abnormalities, clinical experts classify suspicious masses as benign or malignant. This task presents a daily challenge for radiologists due to the huge number of mammograms as well as the time and effort to examine each view of a mammography [4]. Through the use of second opinion or reading by a computer-aided diagnosis (CAD) system, the overall accuracy as well as the false positives and negatives of mass detection, segmentation, and classification could be improved [2]. In the literature, several conventional CAD systems have been developed separately for breast lesion detection, segmentation, or classification [5]. However, there are few studies over a completely integrated CAD system including detection, segmentation, and classification all together. Firstly, detection of breast lesion is an important initial stage to identify the potential region of interest (ROI) of breast lesion in any CAD system. In fact, detection task is still challenging due to the variation of the breast lesions within the surrounding tissues in terms of shape, texture, size, and location in the mammograms [6]. Recently, novel detection approaches based on deep learning were introduced into a CAD system to overcome the challenging tasks of mass detection from mammograms [7]. Secondly, segmentation of breast lesions plays a critical role to accurately extract the specific shape of breast lesions excluding other surrounding normal tissues [8]. Many studies involving mass segmentation have utilized region growing, active contour, and Chan-Vese methods [8]. Unfortunately, these methods still lack performance in handling mass segmentation automatically, because the simple hand-crafted or semi-automatic features based on prior knowledge cannot deal with complex shape variations and different density distribution of the breast lesions. Recently, a few deep learning studies have presented as alternative methodologies for breast lesions segmentations. Indeed, deep learning models have capability to directly extract deep high-level hierarchy feature maps from the input image [8]. Lastly, the majority of CAD systems have been developed to classify the manual extracted breast lesions as either benign or malignant utilizing conventional classifiers [9]. To build such systems, a set of hand-crafted or semi-automatic features describing the characteristics of breast lesions are required. In fact, these conventional CAD systems suffered due to the high degree of similarity of different breast tissues [10]. Alternatively, a few deep learning CAD systems have recently been produced to handle the breast lesion classification task [5]. These systems can learn and extract deep features from input mammograms to achieve better classification performance.

A fully integrated CAD system based on deep learning detection, segmentation, and classification is presented in this chapter. The rest of this chapter is organized as follows. First, You-Only-Look-Once (YOLO) is adopted and used to detect the breast lesions [11]. Second, a newly deep learning model of full resolution convolutional network (FrCN) is produced for breast lesion segmentation. One advantage of FrCN is to preserve the high resolution of feature maps especially at object edges. Finally, three deep learning models including CNN, ResNet-50, and InceptionResNet-V2 are separately adopted and used to distinguish benign and malignant breast lesions. The presented fully integrated CAD system is evaluated using a public INbreast dataset [12] and its performance is compared with the latest deep learning methodologies of detection, segmentation, and classification.

Related Work

Diagnosis of breast cancer via a CAD system could be improved using the capability of deep learning to accurately represent the high-level deep features of breast lesions in mammograms [4]. Detection of breast lesions is an important task to detect the potential abnormalities for any CAD system [4]. Unfortunately, this task is still a big challenging for researchers and has not been fully resolved yet [8]. Previously, manual detection of breast lesions was widely used in building CAD systems even for deep classification CNN [13]. Most of these CAD systems achieved better classification performance against the traditional machine learning techniques [14]. However, the need to automatically detect breast lesions was recently addressed in the several studies [7, 8, 14]. Few studies based on deep learning present automatic methods to detect breast lesions from the input mammograms [8, 14]. Our preliminary detection results of breast lesions via YOLO utilizing the Digital Database for Screening Mammography (DDSM) are presented in [15]. The detection performance via YOLO was better in comparison to the latest deep learning detection methods [8, 11]. In [16], a new deep learning model called region-based CNN (R-CNN) was used to automatically detect the breast lesions [16]. In [8], another automatic method using a very complex cascade of deep learning models was introduced to detect breast lesions based on R-CNN.

For segmentation, several conventional works have been presented to segment the boundaries of breast lesions from the X-ray mammography images such as growing regions, active contour, and Markov random field (MRF) [17]. However, all of these approaches have shortcomings because they required a prior knowledge of breast lesion contours. Recently, few deep learning studies have been presented and achieved better segmentation results for semantic and medical images [8]. These segmentation models are introduced by adopting and converting the classification functionality of VGG-16 to the segmentation purpose. However, these models suffer from the loss of spatial resolution of the generated feature maps due to the multiple layers of max-pooling and subsampling. Although the utilized max-pooling and subsampling layers reduce the dimensions of derived feature maps and minimize the computation expenses, the spatial resolution of those maps are exponentially decreased [5]. In [18], deep learning model of FCN was used to segment skin lesions from dermoscopy images [18]. Inspired by the structure of FCN, another segmentation model called U-Net was introduced to segment neural brain images including encoder and decoder deep convolutional layers [19]. The extracted feature maps from each encoder layer were combined with the corresponding one in the decoder network. Then, decoder up-sampling and deconvolutional layers were performed to overcome the resolution loss of feature maps [19]. In [20], a deep learning segmentation model called SegNet is presented to segment the semantic images. The SegNet model also consisted of encoder and decoder convolutional network stages. Despite the promising segmentation results of these models, they have not yet been applied for breast lesion segmentation. Up to date, only few attempts have been presented for breast lesion segmentation from mammograms based on deep learning.

Breast lesion classification is the last stage of any CAD systems. The aim of this stage is to recognize or classify the breast lesions as either benign or malignant. Indeed, the performance of classification process mainly depends on the efficient representations of the derived features [5]. In 2017, Yu et al. presented that a very deep learning model achieved better classification performance with relatively similar computational cost comparing the shallower models [18]. Recently, few integrated CAD systems based on deep learning have been introduced including detection, segmentation, as well as classification for breast lesions [5, 8, 14]. In [21], a hybrid CAD system was introduced based on the combination of deep and hand-crafted features using CNN-based conditional random forest (CRF) to detect the potential lesion ROIs, region growing based on active contour to segment the lesion boundaries, and CNN to classify the breast lesions. This CAD system achieved diagnosis performance in terms of AUC with 94.10%. In [8, 14], an integrated CAD system for breast cancer detection, segmentation, and classification was presented. For breast lesion detection, a complex cascade structure of deep learning was utilized involving multi-scale deep belief network (DBN) with Gaussian mixture (GM) classifier, two stages of R-CNN, two stages of CRF classifier, and a refinement algorithm based on R-CNN [8]. For segmentation, another cascade of deep learning techniques was utilized involving two stages of DBN-based CRF and a refinement method using Chan-Vese active contour [8]. For classification, a regular feedforward version of CNN was used to classify the breast lesions as either benign or malignant. Despite the successes of these CAD systems for breast cancer diagnosis, the remaining challenges still exist including high complexities of memory, practical implementation, and long prediction time.

Materials and Methods

Our integrated CAD system for breast cancer diagnosis includes detection, segmentation, and classification in a single framework. First, an automatic mass detection based on YOLO is performed. Then, a novel deep learning FrCN segments breast lesions. Finally, an automatic breast lesion classification is performed via the convolutional neural networks. A schematic diagram of the integrated CAD system is depicted in Fig. 1.

Dataset

A public X-ray mammography database, INbreast [12], is used in training and evaluation of our integrated CAD system. The classification label and localization ground truth (GT) of the breast lesions for all mammograms in the INbreast database are available and accurately annotated by the experts [12]. All mammograms were collected to represent real breast data with pixel size of 70 μm (microns), and contrast resolution of 14-bit. According to the breast size of the patient, the mammogram size is 3328 × 4084 or 2560 × 3328 pixels. The INbreast dataset includes 410 mammograms (i.e., normal, benign, and malignant) with both views of MLO and CC

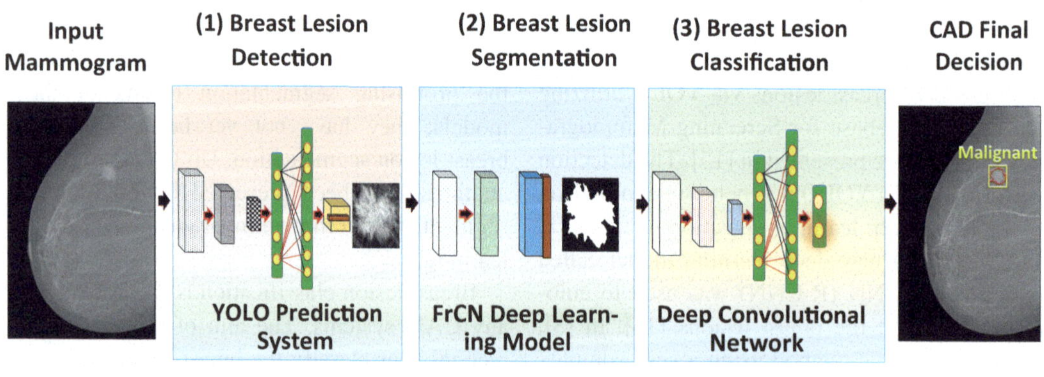

Fig. 1 Schematic diagram of the fully integrated computer-aided diagnosis (CAD) system based on deep learning detection, segmentation, and classification for breast lesions

from 115 patients (cases) [12]. From 90 cases two images (MLO and CC views) are collected for each breast, and from 25 cases only two views (MLO and CC) from one breast having cancer were collected. To evaluate our CAD system, we utilized all mammograms having lesions or masses (benign and malignant) from both views to collect in a total of 107 cases (34 benign and 73 malignant cases) [12]. In INbreast dataset, only two benign and three malignant cases contain two lesions. For benign cases, both CC views are collected from two different patients. For malignant cases, two mammogram views of CC and MLO are collected from the same breast of patient ID: "d713ef5849f98b6c_MG_L," while one CC view is collected from the different patient with ID: "45c7f44839fd9e68_MG_R." Both breast lesions that are visible on both CC and MLO views are used in this study. Thus, 36 benign breast lesions with BI-RAD $\in \{2, 3\}$ and 76 malignant breast lesions with BI-RAD $\in \{4, 5, 6\}$ are collected. Because some of benign and malignant cases have more than one breast lesion, thereby, a total of 112 breast lesions are collected [5, 8, 14].

Datasets Preparation: Training, Validation, and Testing

For data preparation in this work, we randomly divide benign and malignant breast images into three groups: 70% (25 benign and 53 malignant) for training, 10% (4 benign and 8 malignant) for validation, and 20% (7 benign and 15 malignant) for testing as previously performed [8, 14]. In addition, unbiased double cross-validation strategy is utilized as follows. Trainable parameters of the proposed deep learning models are optimized during the training process using only training and validation datasets [14]. Then, the final performance of the presented CAD system is only evaluated using the testing dataset. In fact, double cross-validation is very important for parameters optimization and selections due to the following reasons. First, to be sure testing dataset is totally isolated during the training process. Second, to avoid any bias that may occur during the training process. Third, to ensure that the overall performance of the presented CAD system is robust and reliable for real testing as well. In this study, fivefold cross-validation tests are also carried out with training, validation, and testing sets, which are generated by stratified partitioning to ensure that each breast image gets tested equally and to prevent any bias error for training and testing tasks. For each fold, those breast lesions that are visible on the same mammogram view or on both CC and MLO views from the same patient should be categorized in one set of training, testing, or validation to avoid the system bias as well.

Preprocessing

Preprocessing of all mammograms is achieved using the following steps. First, Otsu's thresholding is used separating the breast region from its background to exclude the unwanted information [5]. Second, contrast limited adaptive histogram equalization (CLAHE) technique has been successfully used to enhance the image contrast between the suspicious lesions and their surrounding normal tissues [5, 8]. Indeed, CLAHE is an image contrast enhancement method which depends on the histogram equalization process [5]. There are two sequential steps to apply CLAHE for breast image enhancement. First, the histogram of the entire mammogram is divided into multi-regions at certain thresholds. Then, the histogram equalization process is locally applied over each region. Same preprocessing strategy is used for all mammograms that used in this study without subsampling.

Data Balancing and Augmentation

To develop deep learning models, a large amount of annotated dataset is required for parameter optimization and selection during training process [7, 8, 14]. To avoid any bias during training process, data balancing and augmentation are widely used as regularization strategies for deep learning models. Indeed, data balancing and augmentation strategy are applied only for training dataset [5, 8, 14]. That means the original breast lesion and its many representations (i.e., augmented data) are only included in the training set. Whereas, the testing set is only used without augmentation to evaluate the proposed CAD system over detection, segmentation, and classification stages. As aforementioned, the training dataset from INbreast is unbalanced containing 25 benign and 53 malignant cases. Balancing of training dataset means generating almost an equal number of breast images from both benign and malignant classes. Due to that, all breast benign mammograms from the INbreast training dataset are vertically flipped to balance benign and malignant cases. Thus, 103 mammograms (i.e., 50 benign and 53 malignant) are generated. Then, the balanced training datasets are augmented 22 times using the following strategies. First, all mammograms are rotated eight times around the origin center with the angle of $\Delta\theta = 45°$ (i.e., 0°, 45°, 90°, 135°, 180°, 225°, 270°, and 315°) [7, 8]. Second, left-right and up-down flipping are applied for all rotated mammograms with 90° and 270°. Third, random scaling and translations are applied ten times for all original breast images. In the total, 2266 mammograms are generated from INbreast to train our CAD system over five-fold cross-validation for all stages of detection, segmentation, and classification. For each k-fold, the strategies of data splitting, balancing, and augmentation are applied in the same ways.

Initialization of Trainable Parameters for Deep Learning Models

To accelerate and avoid overfitting that may occur during training process of all deep learning models, the trainable parameters of convolutional and fully connected (FCs) layers should be initialized [5, 8]. In the literature, several strategies to initialize the trainable parameters of deep learning models were used such as random initialization and transfer learning [5, 8, 14]. In this study, transfer learning is used to initialize the parameters in two consecutive steps. First, all deep learning models are pre-trained using a large annotated computer vision dataset (i.e., ImageNet [22]). Second, these models are fine-tuned using our augmented annotated dataset (i.e., mammograms).

Breast Lesion Detection via YOLO

Detection of breast lesion is the first critical task for the CAD system to identify all potential lesions from entire mammograms. In this chapter, we adopt and use a deep learning model called YOLO, a regional ROI-based CNN technique, to perform the detection task [11]. Our preliminary works using YOLO has proven that this technique is effective for breast lesion detection tasks using a public X-ray mammography dataset, DDSM [5, 15]. Since the INbreast dataset includes accurate ground truth, YOLO could be a good choice for detection of breast lesions due to the following reasons. First, YOLO has a robust ability to directly detect the breast lesions from the entire mammograms [15]. Second, detected bounding boxes via YOLO accurately align the breast lesions, thereby, a low rate of false positives is achieved compared with other studies [14]. Third, YOLO can detect the most challenging cases of breast lesions even when they exist over the pectoral muscles or inside dense regions. Fourth, required testing time and memory are extremely lower than other more complex deep learning models [14].

Breast Lesion Segmentation via FrCN

Once the breast lesions are detected from the previous detection stage of our CAD system, they are directly passed into the novel segmentation model, FrCN, to segment the breast lesions end-to-end. FrCN composes of two main consecutive encoder and decoder networks. The encoder network involves thirteen convolutional layers, while decoder is built using three convolutional layers. Unlike the previous deep learning models, the max-pooling and subsampling layers are removed from the both encoder and decoder networks to preserve the full spatial resolution of the original input image (i.e., breast lesion) as well as the details of the objects. This is a key modification to prevent any information loss during feature map generation. Therefore, the high-level deep feature maps in each block are generated utilizing only the convolutional process, preserving the full

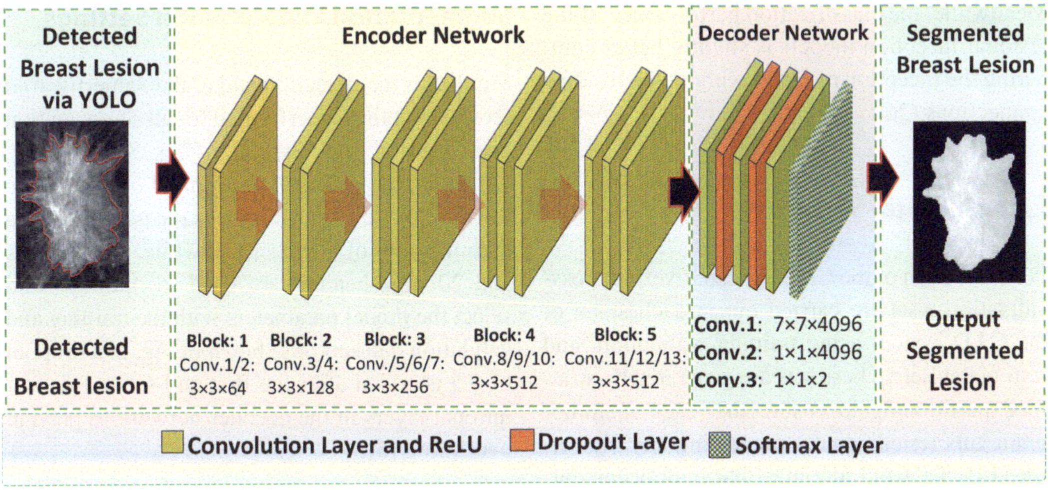

Fig. 2 Segmentation deep learning model of a full resolution convolutional network (FrCN). The input image is a detected breast lesion ROI highlighted with its ground truth (red), while the output image is the corresponding segmented ROI

resolution of the input images. By this modification, FrCN is able to maintain the details and edges especially for the tiny objects. Because the convolutional layers on the full resolution of the input images without subsampling in the encoder network are used, up-sampling and deconvolutional layers in the decoder network are not required. The final output of deep feature maps is directly passed into a softmax function to obtain the probability for each image pixel. Finally, a non-linear activation function of rectified linear unit (ReLU) is utilized after each block in the encoder and decoder networks. The schematic diagram of our deep learning FrCN segmentation model is shown in Fig. 2. To evaluate the overall segmentation performance of FrCN, a direct comparison against other existing deep learning models such as FCN [23], SegNet [20], and U-Net [19] is presented using the same dataset from the INbreast database [12].

Breast Lesion Classification via Three Convolutional Neural Networks

Once breast lesions are detected as well as segmented, deep learning models of our feedforward CNN [5], ResNet-50 [24], and InceptionResNet-V2 [25] are separately used to classify the breast lesions as benign or malignant. These deep learning models are used to perform the classification task of our CAD system. Indeed, we use the regular feedforward CNN presented in our recent study [5]. Also the deep learning models of ResNet-50 and InceptionResNet-V2 are adopted replacing their last two layers by other four layers of global average pooling (GAP) layer, two dense layers with ReLU activation functions, and a logistic regression layer of softmax. Deep learning models such as ResNet-50 and InceptionResNet-V2 are recently introduced with very deep convolutional layers to improve the classification performance preserving the computational burden to be similar as in the shallower CNNs [26]. The main principle of the Inception networks is to produce multiple feature maps from the input images using different parallel pathways with different convolutional filters [26]. By using these remedies with the Inception models, their execution time overcomes other state-of-the-art deep learning models of CNN and ResNet-50 [24]. Since the residual connections are inherently important to train very deep architectures, the filter concatenation stage of Inception models is replaced by the residual ones producing the InceptionResNet models [26]. Training of the Inception models is significantly accelerated using the residual connections as demonstrated in [26].

Meanwhile, the classification performance of the residual Inception models is slightly better comparing the Inception modules without the residual connections [26].

Experimental Settings

For evaluation of the CAD system, fivefold cross-validation tests are carried out in each stage of the CAD system using training, validation, and testing datasets. These sets are generated by stratified partitioning to ensure that each mammogram gets tested equally preventing any bias error [14]. To avoid any bias that may occur during training process, weights optimization process utilizing a weighted cross-entropy loss function as well as double cross-validation are used [5, 14].

Detection Experimental Settings

Due to the different sizes of both mammogram views, all breast images are resized into a fixed size of 448 × 448 pixels [5]. The YOLO-based CAD system is trained for 135 epochs and mini-batch size of 64 using only training and validation datasets. Also, stochastic gradient descent (SGD) optimizer is used with momentum and decay of 0.9 and 0.0005, respectively. The breast lesions are considered to be correctly detected if the intersection over union ($IoU_{GT}^{Ext.}$) between the extracted and ground truth bounding boxes is greater than or equals 50%. Moreover, the false positive candidates of breast lesions are manually excluded before the segmentation and classification stages of the CAD systems as previously applied in [5, 8, 14]. This is because there is a lack of ground truth information of falsely detected lesions to derive the performance evaluation metrics especially for the segmentation stage [8]. Thus, the evaluation results of segmentation and classification tasks are computed with the exception of the falsely detected cases of breast lesions.

Segmentation Experimental Settings

Similar to the detection stage, the same fivefold cross-validation is performed for all segmentation deep learning models: FCN [23], SegNet [20], U-Net [19], and FrCN. To train all of these deep learning models, Adam optimizer is used with a learning rate of 0.001. Meanwhile, 135 epochs and 20 mini-batches are used to optimize and select the model parameters with the training and validation datasets. As shown in Fig. 2, a dropout of 0.5 is added after the first and second convolutional layers in the decoder network to prevent overfitting [5].

Classification Experimental Settings

All detected and segmented breast lesions are normalized and resized using bi-cubic interpolation into a fixed size of 128 × 128, 224 × 224, and 299 × 299 pixels for CNN, ResNet-50, and InceptionResNet-V2, respectively. Then, all these breast lesions are directly fed into the classification stage producing the final prediction of our CAD system. In fact, these deep learning models are adopted to compare the recognition performance of shallower CNN against the deeper models of ResNet-50 and InceptionResNet-V2. This comparison is performed under the same training and testing settings for all deep learning models. To verify the CAD system for classification, the same fivefold cross-validation is performed similar to the detection and segmentation stages. For training, Adam optimizer with the initial learning rate of 0.0001 and weight decay of 0.0005 is used. The learning rate is reduced by 50% if the loss function does not decrease by 0.001 every 10 epochs. The mini-batch size and number of epochs are set to 24 and 130, respectively. Dropout of 0.3 is used for both fully connected layers in all deep learning models to prevent overfitting as well as accelerate the training process [5, 8].

Implementation Environment

All these experiments are performed on a PC with the following hardware specifications: Intel(R) Core (TM) i7-6850K with 16 GB RAM, clock speed of CPU @ 3.360 GHz, and GPU of NVIDIA GeForce GTX 1080. The CAD system is implemented in Python 2.7.14 and C++ on the Ubuntu 16.04 operating system. The implementation of all deep segmentation models is achieved utilizing Theano [27] and Keras [28] deep learning libraries, while the detection and classification models are implemented under the Tensorflow environment [29].

Experimental Results and Discussion

Evaluation Metrics

Each detection, segmentation, and classification stage of the CAD system gets separately evaluated using an overall accuracy (Acc.), sensitivity (Sen.), specificity (Sep.), F1-score (Dice), Jaccard (Jac.), and Matthews correlation coefficient (MCC). Moreover, area under ROC curve (AUC) is also used to evaluate the CAD system over segmentation and classification stages. The definition and criteria for all of these evaluation metrics are available in [5].

Breast Lesion Detection Results

The detection performance of breast lesions via YOLO over fivefold cross-validation tests with the testing dataset is reported in Table 1. The false detection cases presented in Table 1 indicate the cases when $\text{IoU}_{\text{GT}}^{\text{Ext.}} < 50\%$. This means that the final extracted breast lesion has no enough overlap ratio with its GT [5, 7, 8]. Fortunately, YOLO detects at least one bounding box indicating breast lesion from all testing mammograms. In this study, the false detection cases are excluded over each test fold for the next stages of segmentation and classification. An average overall detection accuracy of 97.27% at 0.25 false positive per image (FPI), MCC of 93.93%, and F1-score of 98.02% present the reliable detection performance of the YOLO detector. Examples of the qualitative breast lesion detection results via YOLO identifying the potential breast lesion ROIs are shown in Fig. 3. It is clearly shown that YOLO can accurately detect and align the detected bounding boxes surrounding the breast lesions with high prediction confidence score and high overlapping ratio of IoU. In fact, confidence score indicates the probability of the presence of breast lesions, while overlapping ratio indicates how much the lesion detection localization is accurate. Moreover, a comparison of the detection results using YOLO against the other latest deep learning methods is listed in Table 2. It is clearly shown that YOLO achieved much better detection accuracy with higher prediction speed in comparison to other deep learning detection methods. Also, YOLO can detect even the most challenging cases when the breast lesions exist over the pectoral muscles or inside the breast dense tissues as depicted in Fig. 3(a) and (b), respectively. Therefore, the YOLO detector plays a critical role in the CAD system, achieving the best detection performance of breast lesions comparing the latest deep learning models.

Breast Lesion Segmentation Results

The average segmentation performance results of our FrCN segmentation model against FCN, SegNet, and U-Net over fivefold tests are reported in Table 3. For comparison, the results of all deep learning models are achieved without any refining pre- and/or post-processing. The quantitative measurements of all metrics are computed per pixel of the segmented maps with the same resolution of the input detected breast lesions. FrCN obviously outperformed other methods with an average Dice index of 92.36%, Jaccard coefficient of 85.81%, overall accuracy of 92.69%, and MCC of 85.36%. U-Net achieved better segmentation results comparing SegNet in terms of all evaluation metrics. In addition, SegNet achieved better segmentation performance in terms of specificity with 96.38%.

Table 1 The detection performance of the breast lesions over fivefold cross-validation using YOLO detection model on the test sets from the INbreast dataset

	Benign		Malignant		Total		Metrics (%)		
Fold test	True	False	True	False	True	False	Acc.	MCC	F1-score
1st Fold	6 85.71%	1 14.29%	15 100%	0 0.0%	21 95.45%	1 4.55%	95.45	89.64	96.77
2nd Fold	7 100%	0 0.0%	14 93.33%	1 6.67%	21 95.45%	1 4.55%	95.45	90.37	96.56
3rd Fold	7 100%	0 0.0%	15 100%	0 0.0%	22 100%	0 0.0%	100	100	100
4th Fold	6 85.71%	1 14.29%	15 100%	0 0.0%	21 95.45%	1 4.55%	95.45	89.64	96.77
5th Fold	7 100%	0 0.0%	15 100%	0 0.0%	22 100%	0 0.0%	100	100	100
Avg.(%)	94.28	5.71	98.67	1.33	97.27	2.73	97.27	93.93	98.02

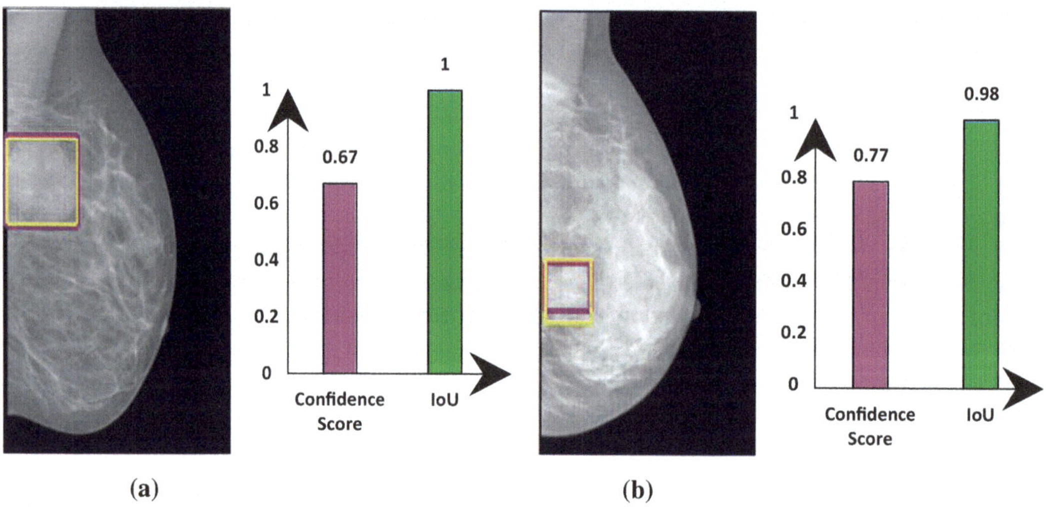

Fig. 3 Examples of detected breast lesions from the INbreast dataset via the YOLO detector. Detected breast lesions exist over the pectoral muscle and inside the breast dense tissues as shown in (**a**) and (**b**), respectively. Detected breast lesions (magenta) and their ground truths (yellow) are superimposed on the original mammograms

Table 2 The detection performance comparison between the YOLO detector against the other latest studies on the test sets from the INbreast

Reference	Method	Dataset	Prediction time per image (Sec.)	Detection accuracy (%)
Dhungel et al. [8], Carneiro et al. [14]	Cascade deep learning F-RCN, DBN, and CRF	INbreast	39	90.0 at 1.0 FPI
Kozegar et al. [30]	Adaptive threshold with some of machine learning techs	INbreast	108	87.0 at 3.67 FPI
Our presented method	YOLO-based CAD system	INbreast	0.014	97.27 at 0.25 FPI

Table 3 The average breast lesion segmentation performance over fivefold cross-validation on the test sets from the INbreast dataset

		Evaluation metrics (%)						
Fold test	Deep learning method	Dice	Jac.	Sen.	Spe.	Acc.	AUC	MCC
Avg. (%)	FCN	88.05	79.14	82.06	95.10	88.95	89.52	80.30
	SegNet	89.40	81.83	83.52	96.83	90.26	90.26	82.04
	U-Net	90.98	83.77	87.03	95.70	91.91	91.72	83.24
	Our FrCN	92.36	85.81	92.94	92.47	92.69	92.70	85.36

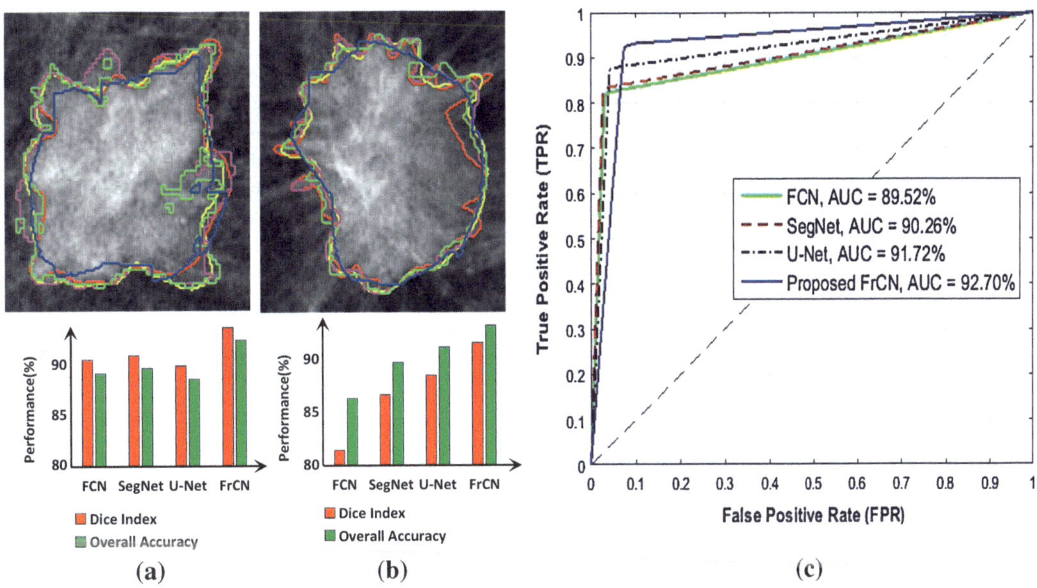

Fig. 4 Examples of qualitatively breast lesion segmentation results via our deep learning model of FrCN against the FCN, U-Net, and SegNet are shown in (**a**) and (**b**), while (**c**) presents the evaluation performance of all deep learning models in terms of ROC curves with their AUCs. The counters in (**a**) and (**b**) indicate the ground truth (red), FrCN (yellow), U-Net (green), SegNet (magenta), and FCN (blue)

Moreover, examples of the qualitative breast lesion segmentation results for FrCN against FCN, SegNet, and U-Net are depicted in Fig. 4 (a) and (b). Moreover, the segmentation performance of FrCN against all other methods is evaluated by the AUC over all test folds. Figure 4(c) shows an example of the ROC curves with AUCs from the 2nd test fold for comparison among all segmentation models. As presented in Table 3 and Fig. 4(c), the performance of breast lesion segmentation via FrCN outperformed all other methods in terms of AUC with 92.70%. In addition, FrCN achieved faster training time with 6.42 h than FCN, SegNet, and U-Net with 12.94, 6.81, and 8.03 h, respectively. For testing, FrCN segments the individual breast lesion in 8.51 s comparing 10.25, 10.48, and 10.66 s for FCN, SegNet, and U-net, respectively. Despite U-Net achieved better segmentation results than SegNet, but it is slightly slower to perform training and testing tasks. FrCN overcomes the limitations of the latest deep learning segmentation models in terms of preserving high resolution and better performance for large and tiny objects.

Breast Lesion Classification Results

Once the detected lesions are segmented via FrCN, deep learning convolutional networks of CNN, ResNet-50, and InceptionResNet-

Table 4 Breast lesion classification performance as an average over fivefold cross-validation on the test sets from the INbreast dataset

		Benign class			Malignant class			Weighted metrics		
Fold test	Method	Sen.	Spe.	Dice	Sen.	Spe.	Dice	ACC	AUC	MCC
Avg. (%)	CNN	84.75	90.57	82.19	90.57	84.76	91.77	88.74	87.67	74.17
	ResNet-50	91.42	93.24	88.35	93.24	91.43	94.52	92.56	92.33	84.10
	Inception ResNet-V2	90.47	97.33	92.16	97.33	90.47	96.64	95.32	93.91	89.39

Table 5 Comparison performance classification results of the fully integrated CAD system against others through all stages of detection, segmentation, and classification

Reference	Prediction classes	Prediction time per image (Sec.)	Overall classification accuracy and (AUC) (%)	Hardware specs
Dhungel et al. [8]	Benign/Malignant	41	91 and (76)	Intel Core i5-2500k, 8GB RAM, 3.30 GHz, and GPU of NVIDIA GeForce GTX 460 SE 4045 MB
Carneiro et al. [14]	Normal/Benign/ Malignant	41	NA and (78) → Benign Vs. Malignant NA and (86) → Malignant Vs. (Normal + Benign)	Intel Core i7, 8GB RAM, 2.3 GHz, and GPU of NVIDIA GeForce GT 650M 1024 MB
Our CAD system	Benign/Malignant	9.32	95.32 and (93.91)	Intel Core i7-6850K, 16GB RAM, 3.360 GHz, and GPU of NVIDIA GeForce GTX 1080

V2 are used to classify these lesions as either benign or malignant. Table 4 shows the average breast lesion classification performance for each class of benign and malignant over fivefold tests. The evaluation results presented in Table 4 are computed for the correctly detected breast lesions. It is obviously noted that the deep learning model of InceptionResNet-V2 achieved better classification results in terms of sensitivity of 97.33%, specificity of 90.47%, overall accuracy of 95.32%, F1-score of 94.40%, and AUC of 93.91%, whereas the shallower model of CNN achieved the lowest classification performance results in comparison to other deeper models of ResNet-50 and InceptionResNet-V2 in terms of all evaluation metrics. This means that the deeper models achieve better classification performance against the shallower ones. However, the promising classification performance of our fully integrated CAD system is achieved due to many reasons as follows. First, the potential breast lesion ROIs are accurately detected and aligned using the prediction model of YOLO. Second, the robust segmentation deep learning model of FrCN plays a critical role to extract the specific region of breast lesions minimizing the false positive and negative pixels from the surrounding normal tissues. Third, the high deep level feature maps derived using the state-of-the-art deep learning models highly contribute to improvement of the overall diagnostic performance of the CAD system. Finally, a comparison between our fully integrated CAD system with respect to the latest studies based on the deep learning is presented in Table 5. All these studies are evaluated using the INbreast dataset through each stage of detection, segmentation, and classification. It is clearly shown that our CAD-based deep learning could handle all these stages achieving a higher performance as well as much faster prediction time. Therefore, the promising prediction performance of the CAD system seems to make it more feasible towards practical applications.

Conclusion

In this chapter, a fully integrated CAD system based on deep learning including detection, segmentation, and classification is presented for automatic diagnosis of the breast lesions in a single framework. To automatically detect breast lesions from the entire mammograms, the YOLO-based lesion detection could be used. Due to the segmentation capability of the FrCN model, the CAD system could achieve a much better diagnostic results. Hence, the detection and segmentation end-to-end of breast lesions could be a key to minimize the false positive and negative rates and then improve the overall performance of our integrated CAD system. Moreover, classification based on the convolutional deep learning contributes to accurately classify breast lesions. A fully integrated CAD system based on deep learning methodologies could be beneficial for practical applications of future medical imaging systems.

Acknowledgements This work was supported by International Collaborative Research and Development Programme funded by the Ministry of Trade, Industry and Energy (MOTIE, Korea) (N0002252). This work was supported by the National Research Foundation of Korea (NRF) grant funded by the Korea government (MEST) (2019R1A2C1003713).

References

1. Siegel RL, Miller KD, Jemal A (2017) Cancer statistics, 2017. CA Cancer J Clin 67(1):7–30
2. Al-antari MA, Al-masni MA, Park SU, Park JH, Kadah YM, Han SM, Kim T-S (2016) Automatic computer-aided diagnosis of breast cancer in digital mammograms via deep belief network. In: Global conference on engineering and applied science (GCEAS), Japan, pp 1306–1314
3. Jill J (2015) Breast cancer screening guidelines in the United States. JAMA 314:1658–1658
4. Al-antari MA, Al-masni MA, Park SU, Park JH, Metwally MK, Kadah YM, Han SM, Kim T-S (2017) An automatic computer-aided diagnosis system for breast cancer in digital mammograms via deep belief network. J Med Biol Eng. https://doi.org/10.1007/s40846-017-0321-6
5. Al-antaria MA, Al-masni MA, Choi M-T, Han S-M (2018) A fully integrated computer-aided diagnosis system for digital X-ray mammograms via deep learning detection, segmentation, and classification. Int J Med Inform 117:44–54
6. Casellas-Grau A, Vives J, Font A, Ochoa C (2016) Positive psychological functioning in breast cancer: an integrative review. Breast 27:136–168
7. Al-Masni MA, Al-Antari MA, Park JM, Gi G, Kim TY, Rivera P, Valarezo E, Choi MT, Han SM, Kim TS (2018) Simultaneous detection and classification of breast masses in digital mammograms via a deep learning YOLO-based CAD system. Comput Methods Programs Biomed 157:85–94
8. Dhungel N, Carneiro G, Bradley AP (2017) A deep learning approach for the analysis of masses in mammograms with minimal user intervention. Med Image Anal 37(1):114–128
9. Yassin NI, Omran S, Houby EM, Allam H (2018) Machine learning techniques for breast cancer computer aided diagnosis using different image modalities: a systematic review. Comput Methods Programs Biomed 156:25–45
10. Chakraborty J, Midya A, Rabidas R (2018) Computer-aided detection and diagnosis of mammographic masses using multi-resolution analysis of oriented tissue patterns. Exp Syst Appl 99(1):168–179
11. Redmon J, Divvala S, Girshick R, Farhadi A (2016) You only look once: unified, real-time object detection. In: 2016 IEEE conference on computer vision and pattern recognition, pp 779–788
12. Moreira I, Amaral I, Domingues I, Cardoso A, Cardoso M, Cardoso J (2012) INbreast: toward a full-field digital mammographic database. Acad Radiol 19(2):236–248
13. Jiao Z, Gao X, Wang Y, Li J (2016) A deep feature based framework for breast masses classification. Neurocomputing 197(C):221–231
14. Carneiro G, Nascimento J, Bradley AP (2017) Automated analysis of unregistered multi-view mammograms with deep learning. IEEE Trans Med Imaging 36(11):2355–2365
15. Al-masni MA, Al-antari MA, Park JM, Gi G, Kim TY, Rivera P, Valarezo E, Han S-M, Kim T-S (2017) Detection and classification of the breast abnormalities in digital mammograms via regional Convolutional Neural Network. In: 39th annual international conference of the IEEE engineering in medicine and biology society (EMBC'17), Jeju Island, South Korea, 2017, pp 1230–1236
16. Ayelet A-B, Karlinsky L, Alpert S, Hasoul S, Ben-Ari R, Barkan E (2016) A region based convolutional net work for tumor detection and classification in breast mammography. In: International workshop on large-scale annotation of biomedical data and expert label synthesis. Springer, Athens, pp 197–205
17. Cardoso JS, Domingues I, Oliveira HP (2015) Closed shortest path in the original coordinates with an application to breast cancer. Int J Pattern Recognit Artif Intell 29(1):2
18. Yuan Y, Chao M, Lo Y-C (2017) Automatic skin lesion segmentation using deep fully convolutional networks with Jaccard distance. IEEE Trans Med Imaging 36(9):1876–1886
19. Ronneberger O, Fischer P, Brox T (2015) U-Net: convolutional networks for biomedical image segmentation. In: International conference on medical image computing and computer-assisted intervention

20. Badrinarayanan V, Kendall A, Cipoll R (2016) SegNet: a deep convolutional encoder-decoder architecture for image segmentation. arXiv preprint arXiv:1511.00561
21. Kooi T, Litjens G, van Ginneken B, Gubern-Mérida A, Sánchez CI, Mann R, den Heeten A, Karssemeijer N (2017) Large scale deep learning for computer aided detection of mammographic lesions. Med Image Anal 35:303–312
22. Krizhevsky A, Sutskever I, Hinton GE (2012) ImageNet classification with deep convolutional neural networks. In: 25th international conference on neural information processing systems, USA, 2012, pp 1097–1105
23. Shelhamer E, Long J, Darrell T (2017) Fully Convolutional networks for semantic segmentation. IEEE Trans Pattern Anal Mach Intell 39(4):640–651
24. He K, Zhang X, Ren S, Sun J (2015) Deep residual learning for image recognition. arXiv:1602.07261v2, pp 770–787
25. Simonyan K, Zisserman A (2014) Very deep convolutional networks for large-scale image recognition. arXiv preprintarXiv:1409.1556
26. Szegedy V, Ioffe S, Vanhoucke V (2016) Inception-v4, Inception-ResNet and the impact of residual connections on learning. arXiv:1602.07261v2 [cs.CV]
27. Lab L (2017) Theano. University of Montreal [Online]. Available: http://deeplearning.net/software/theano/tutorial/. Accessed 10 2017
28. Chollet F (2017) Keras: the Python deep learning library. MIT, [Online]. Available: https://keras.io/. Accessed 10 2017
29. Google Brain Team, TensorFlow, 9 11 2017. [Online]. Available: www.tensorflow.org. Accessed 10 2017
30. Kozegar E, Soryani M, Minaei B, Inês D (2013) Assessment of a novel mass detection algorithm in mammograms. J Cancer Res Ther:592–600

Decision Support System for Lung Cancer Using PET/CT and Microscopic Images

Atsushi Teramoto, Ayumi Yamada, Tetsuya Tsukamoto, Kazuyoshi Imaizumi, Hiroshi Toyama, Kuniaki Saito, and Hiroshi Fujita

Abstract

Lung cancer is the most common cancer among men and the third most common among women in the world. Many diagnostic techniques have been introduced to diagnose lung cancer. Positron emission tomography (PET)/computed tomography (CT) examination is an image diagnostic method that performs automatic detection and distinction of lung lesions. In addition, pathological examination by biopsy is performed for lesions that are suspected of being malignant, and appropriate treatment methods are applied according to the diagnosis results. Currently, lung cancer diagnosis is performed through coordination between respiratory, radiation, and pathological diagnosis experts, but there are some tasks, such as image diagnosis, that require a large amount of time and effort to complete. Therefore, we developed a decision support system using PET/CT and microscopic images at the time of image diagnosis, which leads to appropriate treatment. In this chapter, we introduce the proposed system using deep learning and radiomic techniques.

Keywords

Lung cancer · Nodule · Detection · Classification · PET/CT · CT · Cytology · Deep learning · Convolutional neural network

Introduction

Worldwide, lung cancer is the most common cancer among men and the third most common among women [1]. In 2018 alone, there were two million new cases diagnosed. Currently, lung cancer is the leading cause of death among mankind in the USA, Europe, and many Asian countries, making it a serious health problem across the world.

Currently, computed tomography (CT) is widely used for lung cancer screening [2]. According to the results of a national lung

A. Teramoto (✉) · A. Yamada · K. Saito
Faculty of Radiological Technology, School of Medical Sciences, Fujita Health University, Toyoake, Japan
e-mail: teramoto@fujita-hu.ac.jp; ayumi926@fujita-hu.ac.jp; saitok@fujita-hu.ac.jp

T. Tsukamoto · K. Imaizumi · H. Toyama
School of Medicine, Fujita Health University, Toyoake, Japan
e-mail: ttsukamt@fujita-hu.ac.jp; jeanluc@fujita-hu.ac.jp; htoyama@fujita-hu.ac.jp

H. Fujita
Faculty of Engineering, Gifu University, Gifu, Japan
e-mail: fujita@fjt.info.gifu-u.ac.jp

G. Lee, H. Fujita (eds.), *Deep Learning in Medical Image Analysis*, Advances in Experimental Medicine and Biology 1213, https://doi.org/10.1007/978-3-030-33128-3_5

screening trial [3], low-dose CT scan-based screening reduces deaths by 20%, and thus it is regarded as a suitable diagnostic tool for the early detection of lung cancer. Recently, in some countries, positron emission tomography (PET)/CT has also been adopted as a mass screening tool for the diagnoses of cancers [4, 5]. In this combined technique, PET images provide functional information, while CT images render anatomical information, making it possible to detect pulmonary nodules precisely and to classify between benign and malignant nodules. Wever et al. reported on the clinical effectiveness of PET/CT in detection and characterization [6]. In the near future, PET/CT examination will be widely used for screening.

If a suspicious lesion is found by radiological examinations, a subsequent pathological diagnosis is performed to check the lesion in detail [7]. Cytological diagnosis and histological diagnosis are conducted to determine whether it is benign or malignant and to determine cancer type using a specimen made from tissue, respectively. The major tissue types of lung cancer are adenocarcinoma, squamous cell carcinoma, small cell carcinoma, and large cell carcinoma in that order. Other than small cell carcinoma, these types are often treated in the same manner. Surgery is performed for non-small cell carcinomas of removable size, and small cell carcinomas with high metastases are selected for drug therapy and radiation therapy. Recently, molecularly targeted drugs and immunotherapies have been added to drug treatment options and have achieved good therapeutic results. For these two therapies, genetic testing is performed to determine the genotype of cancer cells, and appropriate drugs are selected.

Therefore, in the current diagnosis and treatment of lung cancer, early detection of a lesion and accurate analysis of the lesion can lead to correct treatment and improve the prognosis of the patient. Currently, lung cancer diagnosis is performed through coordination between respiratory, radiation, and pathological diagnosis experts, but there are some tasks, such as image diagnosis, that require a large amount of time and effort to complete. Therefore, we developed a decision support system using PET/CT and cytological images at the time of image diagnosis, which leads to appropriate treatment.

In this chapter, we introduce some of our work. First, an outline of the decision support system for lung cancer diagnosis is given in section "Outline of Decision Support System." In sections "Automated Detection of Lung Nodules in PET/CT Images Using Convolutional Neural Network and Radiomic Features" and "Automated Malignancy Analysis of Lung Nodules in PET/CT Images Using Radiomic Features," automatic detection of lung nodules using PET/CT images and benign/malignant discrimination methods are introduced. Classifications of benign/malignant cells and lung cancer type are described in sections "Automated Malignancy Analysis Using Lung Cytological Images" and "Automated Classification of Lung Cancer Types from Cytological Images." Finally, this chapter is concluded in section "Conclusion."

Outline of Decision Support System

Figure 1 illustrates the concept of the decision-making system for lung cancer diagnosis. Automatic detection of nodules is performed using two

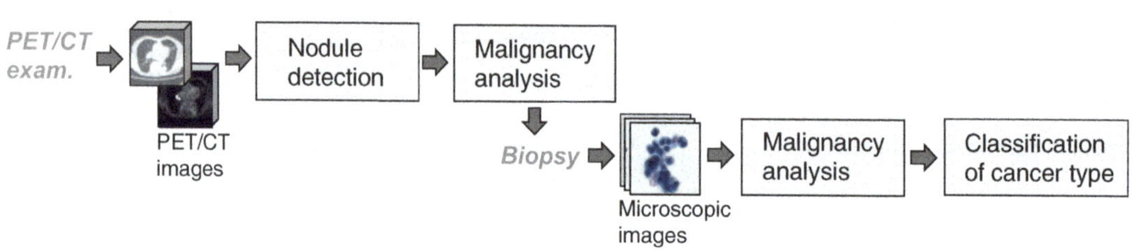

Fig. 1 Schematic diagram of decision support system for lung cancer

types of images obtained by PET/CT examination, and benign and malignant detected nodules are then classified automatically. If the result is malignant, microscopic images are obtained via pathological examination. After analyzing the benign and malignant cell images, classification of lung cancer type is performed for malignant cells. Physician confirmation is made on the output of each automated scheme, and a final diagnosis is made.

Automated Detection of Lung Nodules in PET/CT Images Using Convolutional Neural Network and Radiomic Features

Background

Lung cancer is the leading cause of cancer-related deaths among men [1]. Recently, PET/CT examination is currently being used as a cancer screening tool [4, 5] in some countries. PET/CT is useful for the early detection of lung cancer. However, interpretation using a large amount of PET/CT images is burdensome, and a technique for obtaining high diagnostic accuracy with little effort is required.

We are currently developing a method for the automatic detection of lung nodules from PET/CT images [8–10]. In our previous studies, nodules were detected independently from CT and PET images, and they were integrated to obtain initial candidate regions. Then, handcrafted features were extracted from them and false positives (FPs) were deleted using a support vector machine (SVM). Nodule detection ability was 9.8 false positives/case when the detection sensitivity was 90% [9], and it was a challenge to improve FP reduction performance.

As a method of classifying false positives and nodules, we focused on the convolutional neural network (CNN), which is an artificial intelligence technology with high performance in image recognition tasks. In this section, we will introduce the novel scheme of lung nodule detection from PET/CT images using a CNN and radiomic features [11].

Method Overview

An overview of the detection of lung nodules is shown in Fig. 2. First, initial nodule candidates are detected separately from the PET and CT images using algorithms specific to each image type. Then, candidate regions obtained from the two images are combined. FPs contained in the initial candidates are eliminated by an ensemble method using multistep classifiers on radiomic features obtained by a shape/metabolic analysis and a CNN.

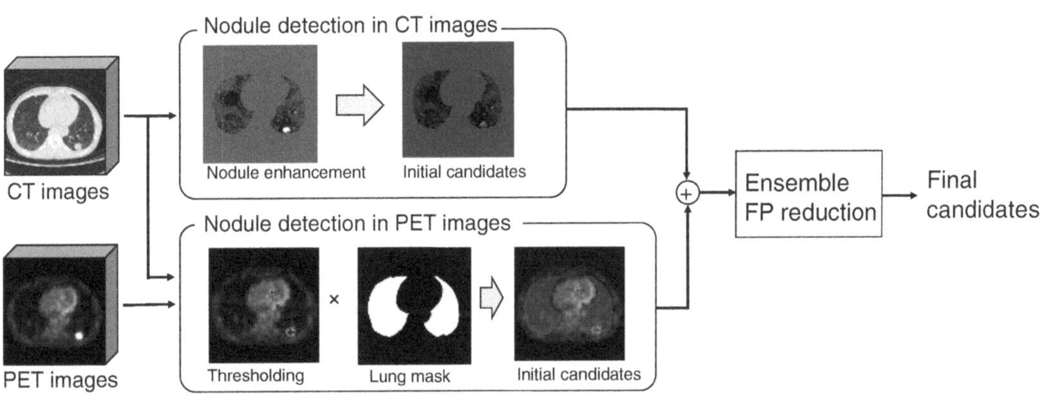

Fig. 2 Outline of nodule detection scheme

Initial Nodule Detection

With regard to the detection in CT images, the massive region was first enhanced using an active contour filter (ACF) [9], which is a type of contrast-enhancement filter with a deformable kernel shape, as shown in Fig. 3. The active contour involves several nodes that are connected to each other. We define the evaluation function of the active contour as the maximum pixel value along the connected lines. The nodes move iteratively to minimize the evaluation function. Thus, the active contour encloses the nodule without touching normal organs, such as blood vessels and the lung wall. The final output of the ACF is the difference between the maximum pixel value on the active contour and the pixel value at the center of the filter kernel. Subsequent to image enhancement, the initial nodule regions are segmented by thresholding and labeling (top of Fig. 2).

The PET images are subsequently binarized using a predetermined cut-off value to detect regions of increased uptake. Here, candidate regions other than the lungs are eliminated using the lung regions obtained by CT images (bottom of Fig. 2).

Initial candidate regions detected by CT and PET are represented as binary images. The two images are then combined using the logical OR function. Following pixel-by-pixel confirmation of regions on both images, a region detected by at least one modality is treated as an initial candidate region.

False Positive Reduction

Among the initial candidate regions, there is a large number of FPs. FPs included in the initial candidates are composed of bronchi and blood vessels in the lung regions. In addition, most of the FPs of the initial candidates in PET images are due to physiological uptakes in organs adjacent to the lungs. Therefore, the integration of both shape features from CT images and metabolic features from PET images can be used to eliminate FPs [8]. However, the features of FPs and nodules overlap, and there is a limit in FP reduction using only these handcrafted radiomic features. To eliminate FPs while maintaining the true positive (TP) rate, this study employed a CNN, which is a type of deep learning architecture [12]. The CNN was inspired by biological processes and specifically designed to emulate the behavior of visual systems. CNNs have a function that automatically performs feature extraction from image data and learns the representation of input data. In some image recognition trials, results were dramatically improved by employing CNNs [13, 14]. Studies have indicated that CNNs can be used to reduce FPs by generating novel valid features not generated by the shape and metabolic features used in conventional FP reduction methods. Therefore, the ensemble FP reduction method was developed for this study by combining a CNN with our previous FP reduction technique, which uses handcrafted radiomic features, as illustrated in Fig. 4.

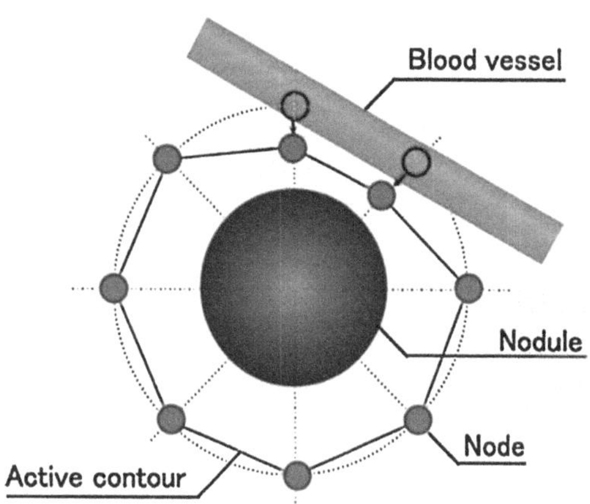

Fig. 3 Active contour filter for nodule enhancement

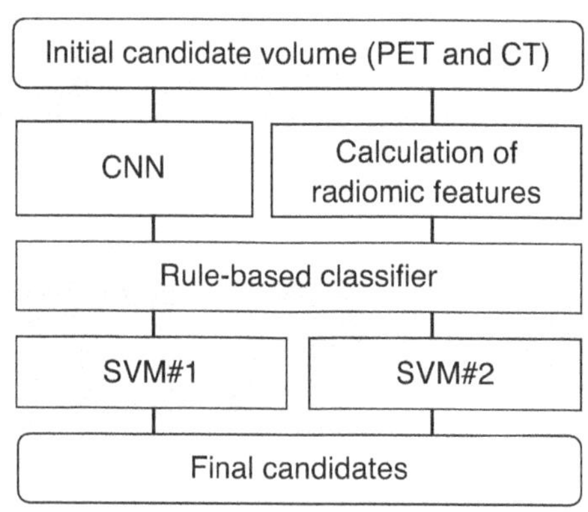

Fig. 4 Flowchart of ensemble FP reduction

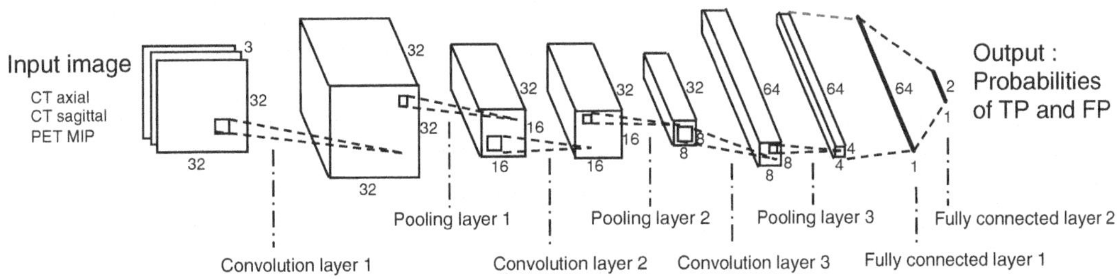

Fig. 5 Architecture of convolutional neural network (CNN)

Classification Using a Convolutional Neural Network

The architecture of the CNN used for FP reduction is illustrated in Fig. 5. It is composed of three convolution layers, three pooling layers, and two fully connected layers. The three sectional CT and PET images are fed to the input layer of the CNN. Axial and sagittal images are then introduced for the CT images. Because PET images provide limited anatomical information and low spatial resolution, it is more important to identify the existence of high uptake regions than the three-dimensional structure. Therefore, we used a maximum-intensity projection (MIP) image along the body axis of the PET images. The input to the first convolutional layer is $32 \times 32 \times 3$ images, where CT-axial, CT-sagittal, and PET-MIP images are each resized to 32×32 pixels.

Convolution layer 1 uses 32 filters with a $5 \times 5 \times 3$ kernel, resulting in a feature map of $32 \times 32 \times 32$ pixels. Pooling layer 1 conducts subsampling, which outputs the maximum value in the 3×3 kernel for every 2 pixels, reducing the matrix size of the feature map to $16 \times 16 \times 32$. After the three convolution layers and three pooling layers, there are two fully connected layers, each consisting of a multi-layer perceptron. After all layers are completed, the probabilities of TP and FP are obtained from the output. By training the convolution and fully connected layers, two separate outputs can represent the probabilities of judgment for FP and TP.

Handcrafted Radiomic Features

For each candidate region, shape and metabolic features are calculated as handcrafted radiomic features. A total of 18 features are obtained from the CT images, including sectional areas in the three planes (X-Y, X-Z, and Y-Z), volume, surface area, contour pixels in the three planes, compactness, convergence in the three planes, and CT values (max, center, and standard deviation) in the candidate region [8]. A total of 8 metabolic features are obtained from the PET images, including the standardized uptake value (SUV) [15] at the center of the candidate region, the maximum and mean values of SUV in the candidate region, the sectional areas in the three planes, volume, and surface area in the candidate region.

Classification

Using the CNN output and handcrafted radiomic features, FPs are eliminated from the initial candidate region. First, the handcrafted radiomic features are fed to the rule-based classifier to eliminate the obvious FPs [8]. FPs are then identified by setting simple low and high limits for each feature.

The remaining candidate regions are then fed to the two SVMs, where the TPs and FPs are classified as shown in Fig. 4. The initial candidate regions detected only by CT images indicate that there are no high uptake regions in the PET images. Therefore, many features obtained by PET are set to zero. In contrast, for the initial candidates detected by PET images, morphological changes are usually observed in CT images.

Because the properties of the obtained features and the number of effective features are different under the two conditions, two SVMs were introduced. A total of 22 features (including all features obtained by CT, three SUV features by PET, and the CNN output) are fed to the first SVM, which is referred to as SVM #1, and all features obtained by the proposed method are fed to the second SVM, which is referred to as SVM #2, based on the above discussion. In this study, LIBSVM, which is a library of SVMs, was introduced [16]. We used C-support vector classification as an SVM algorithm and the radial basis function as a kernel function.

Results

Image Datasets

A total of 104 Japanese individuals who underwent whole-body PET/CT during cancer screening programs from 2009 to 2012 were included in this study. The images were acquired during a cancer screening program at the East Nagoya Imaging Diagnosis Center (Nagoya, Japan) using a Siemens TruePoint Biograph 40 with standard clinical settings. The spatial resolution of the PET images was $4.0 \times 4.0 \times 2.0$ mm^3, while that of the CT images was $0.97 \times 0.97 \times 2.0$ mm^3. A total of 183 nodules were detected in 84 patients. The average values of diameter, CT value, and SUV max for these nodules were 18.9 ± 15.6 mm, 25.3 ± 384.4, and 4.01 ± 4.70, respectively. The center coordinate (x-y-z) of nodules was provided by the radiologist.

This study was approved by an institutional review board, and patient agreements were obtained under the condition that all data would be anonymized.

Evaluation Metrics

The data pertaining to candidate regions were randomly divided into five sets and evaluated using the cross-validation method. A candidate nodule was considered correctly detected if the center coordinates of the nodule marked by a doctor existed inside the candidate region (area) obtained by the proposed method. An initial candidate region was considered to be an FP when no registered nodules were assigned to the region. With regard to the detection parameters, the maximum filter radius of ACF was set at 25 mm, while the number of nodes was set at 8. For detection with PET, the threshold was set at 2.0. These parameters were determined in a previous study, which is based on preliminary experiments and the knowledge of radiologists.

The automated detection calculation was performed using in-house CAD software using an Intel Core i7-6700K processor (4 CPU cores, 4 GHz) with 16 GB of DDR4 memory. For the ensemble FP reduction method, the training of the CNN was conducted using a dedicated training program bundled in the Caffe package [17], which is accelerated by a GPU (NVIDIA GeForce GTX 970 with 4 GB of memory).

Detection Results

In the initial detection, among 181 nodules, 163 and 80 were detected through CT images and PET images, respectively. Among these detected nodules, 67 nodules were detected through both images; total sensitivity was 97.2%. In addition, 7575 FPs were contained in the initial candidates, so the number of FPs/case was 72.8.

The free-response receiver operating characteristic (FROC) curves made by changing the parameters of the FP reduction method are shown in Fig. 6. To conduct a comparative evaluation, the FROC curve of the previous method, which does not employ a CNN, is also shown [9]. The sensitivity of our proposed method was 90.1% with 4.9 FPs/case. Examples of nodules detected or missed by our proposed method are shown in Fig. 7. Figure 7(a) shows three nodules detected by both our previous and proposed methods; Fig. 7(b) indicates three nodules newly detected by this method, and Fig. 7(c) shows three nodules missed by our proposed method. In Fig. 7(c), the left and center images are of nodules missed at the initial detection stage, and the right one is of a nodule missed at the FP reduction stage.

In the present system, initial detection of nodules from CT and PET images was performed automatically by in-house software. There were no manual interactions involved that required judgment of the operator. The processing time for initial nodule detection was approximately 340 s

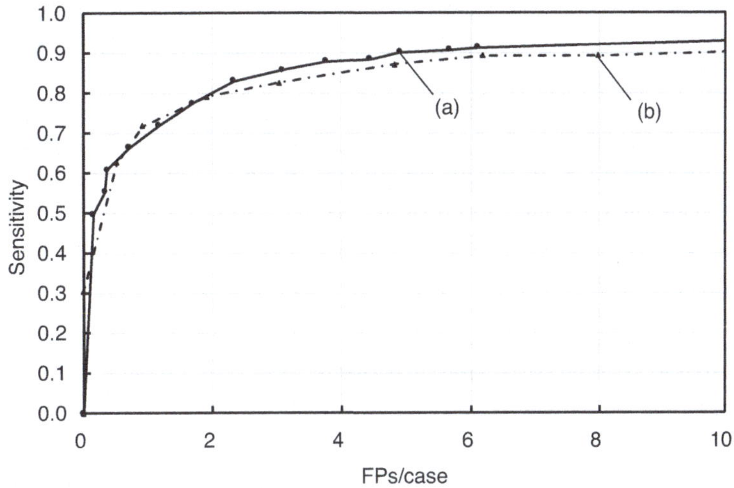

Fig. 6 Free-response receiver operating characteristic (FROC) curves of the (**a**) proposed and (**b**) previous methods

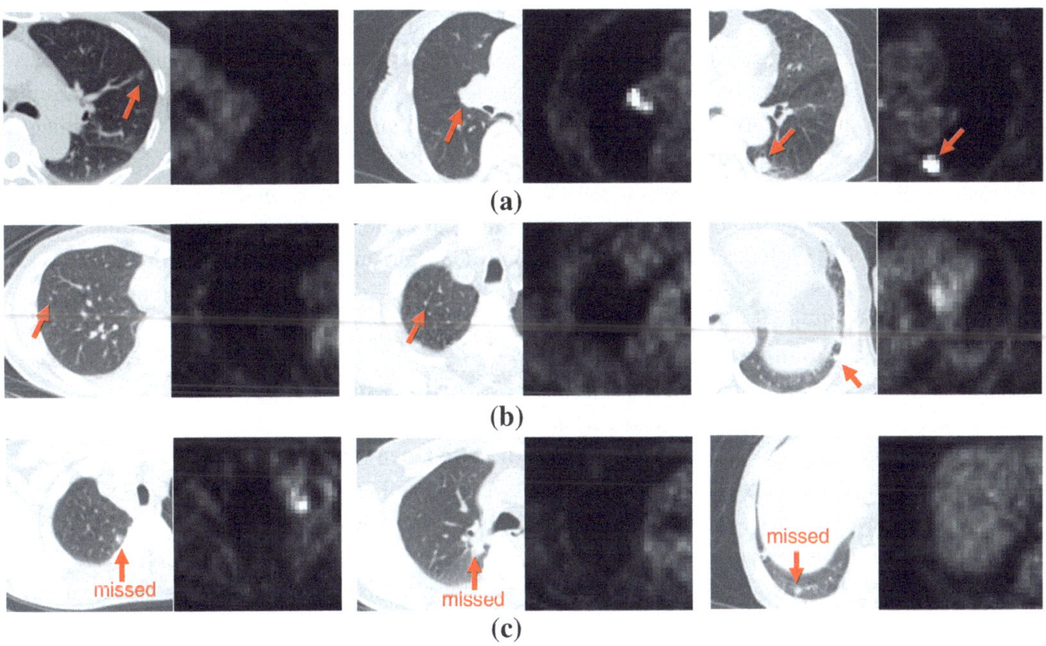

Fig. 7 Examples of detected and missed nodules in pairs of transverse views of CT (left) and PET (right) images

per case, and ensemble FP reduction took 35 s per case. The CNN training took approximately 24 min on the GPU.

Discussion

In this section, we describe the nodule detection method for PET/CT images using a CNN and conventional handcrafted features. In terms of the overall performance using the ensemble FP reduction technique with a CNN and radiomic features as shown in Fig. 6, sensitivity exceeded 90% with 4.9 FPs/case. The proposed CAD scheme therefore appears to be useful for image diagnosis by radiologists during screening and follow-up examinations.

Comparing the proposed method and conventional methods in terms of FPs/case at the same detection sensitivity, our previously proposed method [9] without a CNN had 9.8 FPs/case. Our ensemble FP reduction method using the CNN technique eliminates approximately half the FPs existing in the previous method. In a state that has sufficiently high detection sensitivity, this improvement is noteworthy. These results indicate that our method may be useful for computer-aided detection of lung tumors in clinical practice.

Automated Malignancy Analysis of Lung Nodules in PET/CT Images Using Radiomic Features

Introduction

PET/CT examination is performed for detailed analysis of lung disease. In this combined technique, PET images provide functional information, while CT images render anatomical information, thereby facilitating a comprehensive analysis of the malignancy of nodules.

However, fluorodeoxyglucose (FDG) PET images of benign nodules, such as those associated with inflammatory diseases, often exhibit high uptake values similar to those of malignant nodules. Furthermore, they are anatomically similar in structure. Therefore, it is often difficult to differentiate between benign and malignant nodules [18]. In such cases, a bronchoscopic biopsy is performed; however, this is invasive, and the patient faces great physical hardship.

In these cases, if the CT and PET images can be analyzed in detail to quantify the degree of malignancy of the nodules, the need for an excessive biopsy with its accompanying physical hardship can be reduced. Therefore, in this study, we focused on the automated analysis of the malignant potential of lung nodules using PET/CT images.

Many studies have reported on benign/malignant differentiation of lung nodules by image analysis [19–24]. Armato et al. proposed the automated analysis of lung nodules using linear discriminant analysis with characteristic features obtained from CT images [19]. They evaluated 470 CT scans and revealed that the area under the ROC curve was 0.79. Shen et al. introduced a deep learning multi-crop convolutional neural network model to classify lung nodules [23]. They used 880 benign nodules and 495 malignant nodules from the Lung Image Database Consortium and Image Database Resource Initiative (LIDC/IDRI) dataset and achieved an accuracy of 87.14%. Nie et al. developed a semi-automated scheme for distinguishing between benign and malignant lung nodules by integrating PET and CT information [24]. The study evaluated three computer-aided diagnosis schemes based on an artificial neural network to distinguish between benign and lung nodules using clinical information and image features; it was seen that diagnostic accuracy using both PET and CT was better than using either alone.

However, to the best of our knowledge, automated analysis of lung nodules based on radiomic features obtained from both PET and CT images has not been developed so far. The automated classification of lung nodules could have great practical value. Therefore, in this section, we will introduce an automated classification scheme for lung nodules using CT and PET images [25].

Materials and Methods

Image Dataset

We collected 36 early and delayed phase PET/CT images from patients with suspected lung cancer diagnoses. In addition, conventional CT images at maximal inspiration were also analyzed. The cases chosen were those where differential diagnosis was difficult with diagnostic imaging alone, and the final diagnosis was made by bronchoscopy and biopsy specimen analysis.

The PET/CT imaging studies were performed at the Fujita Health University Hospital (Toyoake, Japan) using Siemens True Point mCT from 2012 to 2014. Both images were obtained with a matrix size of 200 × 200 pixels (voxel size was 4.07 × 4.07 × 2.00 mm^3 and scan time was 2.0 min/table) and under free breathing. Image reconstruction was performed using the 3D-OSEM reconstruction algorithm. In addition, the point-spread function and time-of-flight correction (PSF + TOF) and attenuation correction by CT images were performed. PET images were converted to the voxel size of the CT images after reconstruction. Early and delayed PET imaging was performed 60 min and 120 min, respectively, after the administration of 3.7 MBq/kg of FDG. These PET and CT images were aligned automatically by the PET/CT scanner.

Conventional CT imaging was performed using Aquilion ONE (Toshiba) with a matrix size of 512 × 512 pixels (voxel size was 0.625 × 0.625 × 0.500 mm^3) with the lung kernel. If CT examination was performed more than once, we selected images taken at the time closest to the PET/CT examination.

Of the total 18 benign cases, 13 were finally diagnosed by biopsy, and 5 cases were confirmed to be benign by follow-up examinations over 3 years or more. There were 18 malignant cases, of which there were 13 cases of adenocarcinoma, 4 cases of squamous cell carcinoma, and one case of small cell carcinoma. The age of a patient in the malignant nodule group was 72.2 ± 7.6 years, and in the benign nodule group it was 65.3 ± 10.2 years.

This study was approved by the institutional review board, and patient agreements were obtained under the condition that all data would be anonymized (No. HM17–002).

Methods Overview

An overview of the proposed method is illustrated in Fig. 8. In this method, regions of PET and CT images designated as suspicious by the doctor were analyzed using several characteristic features and were automatically classified and identified as benign or malignant.

Volume of Interest (VOI) Extraction

The position and diameter of the nodule to be analyzed using conventional CT images and the PET/CT images were specified by the physician. Based on that information, segmentation of the volume of interest (VOI) around the lung nodule from the CT and PET images was conducted and employed for analysis. The center coordinates of the VOIs extracted from conventional CT and PET/CT images were manually set while checking the multiplanar reconstruction (MPR) image of each CT image. First, the trans-axial image with the largest nodule area was searched, and its center coordinates were specified manually. Then, we set the longest diameter in the image as D_{xy} in the x-y direction (trans-axial plane) of the nodule. Subsequently, while changing the slice position in the body axis direction, we obtained the range of the slice in which the nodule exists and set it as D_z. We extracted the VOI whose number of pixels on each side was $2D_{xy}$, $2D_{xy}$, and $2D_z$ from the original image.

Extraction of Characteristic Features

Checkpoints in Malignancy Diagnosis

Table 1 lists the checkpoints for distinguishing between benign and malignant nodules [26, 27]. The physicians created this scheme by referring to the pixel values, such as the uptake value of the PET images and the CT values. Furthermore, consideration was given to nodule components

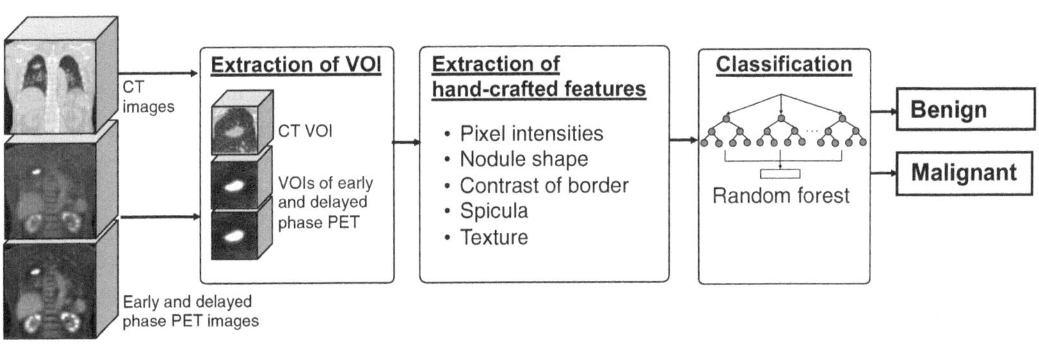

Fig. 8 Overview of nodule classification method

Table 1 Checkpoints of features in benign and malignant nodules

	FDG uptake	CT value	Component	Shape	Border	Spicule
Benign	Low	Low	GGO	Line-like	Clear	Few
Malignancy	High increase at delayed phase	High	Solid	Ball-like	Unclear	Many

(GGO or solid), shapes (roundness), clarity of the nodule border, spicules, etc. In this study, these points were quantified as characteristic features.

Features Regarding Pixel Intensities of PET and CT Images

Many malignant nodules have high pixel intensity in PET and CT images. Therefore, the SUVs [15] of early and delayed PET images were defined as *ESUV* and *DSUV*, respectively. Furthermore, the difference in SUV between the delayed phase and early phase was defined as ΔSUV. In the measurement of SUV, we introduced two methods: SUV_{max} (hottest voxel) and SUV_{peak} (maximum average SUV within a 1 cm^3 spherical volume). In the CT images, the CT value at the center of the nodule (CT_{centre}) and the maximum CT value inside the nodule (CT_{max}) were calculated.

Features Regarding Nodule Shape

Malignant nodules often have a ball-like shape, and benign nodules have a line-like shape. To evaluate the ball-like and line-like shapes, a Hessian matrix method was proposed [28]. The Hessian matrix was obtained by taking the second order differential of the three-dimensional image as follows:

$$H = \begin{bmatrix} F_{xx} & F_{xy} & F_{xz} \\ F_{yx} & F_{yy} & F_{yz} \\ F_{zx} & F_{zy} & F_{zz} \end{bmatrix}. \quad (1)$$

Then, three eigenvalues (λ_1, λ_2, λ_3) were obtained from the matrix. Finally, the ball-like and line-like features (L_{mass} and L_{line}) were calculated using the eigenvalues as follows:

$$L_{mass} = |\lambda_3| / \lambda_1 \quad (2)$$

$$L_{line} = |\lambda_2| (\lambda_2 - \lambda_3) / |\lambda_1| . \quad (3)$$

Features Regarding Contrast of the Nodule Border

The border of a malignant nodule is often unclear. Therefore, the contrast of borders was evaluated using the difference between the CT values of the outer and inner border lines of the nodules. To calculate this, the average CT values at the pixels belonging to the inner edge R1 (CT_{R1}) and the peripheral region R2 (CT_{R2}) were obtained, and the difference between the two values $|CT_{R1} - CT_{R2}|$ was defined as the contrast C_b (Fig. 9(a)). To obtain R1 and R2, the image was first binarized and the contour was extracted by the Sobel operator. Then, the set of pixels on the outline was defined as R2. Next, the binarized region was shrunk by a morphological operation (erosion) with a structural element having a radius of 1 pixel, and the contour of the reduced region was extracted in the same manner as described above; the set of these pixels was used as R1.

Features Regarding Spicules

When there is a spicule around a nodule, it is more likely to be malignant. In this study, spicules in CT images were detected using a Gabor filter [29, 30]. By applying the Gabor filter, line patterns and their orientations were obtained (Fig. 9 (b-2) and (b-3)). From the two images, radial line patterns were extracted (Fig. 9 (b-4)), and the numbers of radial components and their ratios were calculated as the features of spicules SP_1 and SP_2.

Texture Features

The texture pattern of the lung lesion is important for evaluating malignancy. Textures can be analyzed in several ways. We used a method based on the gray level co-occurrence matrix (GLCM) proposed by Haralick et al. [31]. The matrix element $P(i,j)$ of GLCM is the set of second order statistical probability values for changes between gray levels i and j at a particular distance d and angle θ, where θ represents the counter-clockwise angle

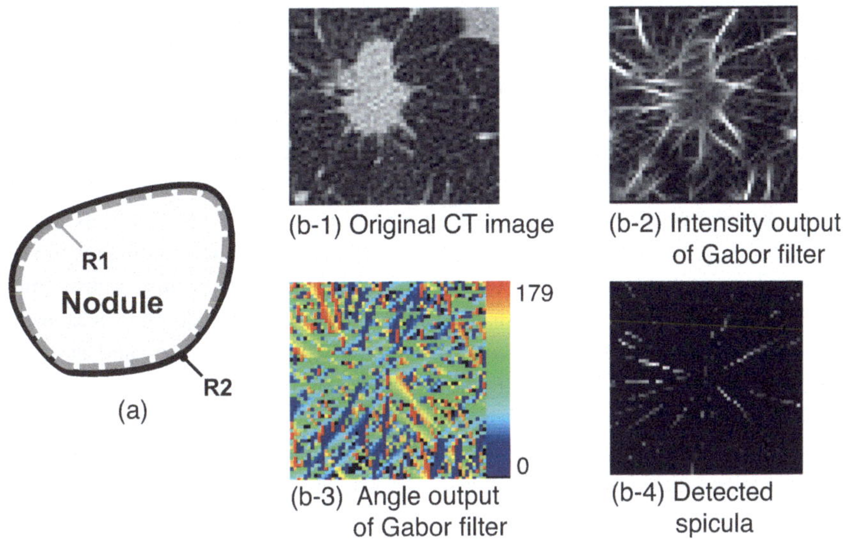

Fig. 9 Handcrafted features

from the X axis. The GLCM can be used to assess properties such as texture uniformity, directionality, and contrast based on the distribution of the values of the matrix elements. Haralick et al. proposed 14 characteristic features using GLCM. In this study, it was necessary to limit the number of characteristic features because the number of cases collected was small. To obtain texture features concerning direction, we calculated the following five features $T_1 - T_5$ using θ values of 0 ($T_{1\_0} - T_{5\_0}$) and 90 degrees ($T_{1\_90} - T_{5\_90}$) in the trans-axial plane of CT images:

- Contrast

$$T_1 = \sum_i \sum_j P(i, j)\ |i - j|^2 \tag{4}$$

- Dissimilarity

$$T_2 = \sum_i \sum_j P(i, j)\,|i - j| \tag{5}$$

- Correlation

$$T_3 = \sum_i \sum_j (i - \mu_i)\left(j - \mu_j\right) P(i, j)\,/\sigma_i \sigma_j, \tag{6}$$

where

$$\mu_i = \sum_i \sum_j i P(i, j)\,, \mu_j = \sum_i \sum_j j P(i, j) \tag{7}$$

$$\sigma_i = \sqrt{\sum_i \sum_j P(i, j)\,(i - \mu_i)^2},$$
$$\sigma_j = \sqrt{\sum_i \sum_j P(i, j)\left(j - \mu_j\right)^2} \tag{8}$$

- Homogeneity

$$T_4 = \sum_i \sum_j P(i, j)\,/\,(1 + |i - j|) \tag{9}$$

- Energy

$$T_5 = \sqrt{\sum_i \sum_j P(i, j)^2} \tag{10}$$

Classification

Identification as benign or malignant was conducted using the obtained characteristic features. In this study, we used the random forest algorithm for classification [32]. The random forest is an ensemble learning method for classification and regression that operates by constructing multiple decision trees and outputting the class that is the mode of the classes of the individual trees. Practically, the input for the random forest was the 25 characteristic values, and a judgment result (benign or malignancy) was obtained from the output. In this study, the maximum number of trees was set to 20.

In addition, to analyze the distinguishing characteristics of this method, we evaluated three kinds of classification methods: (1) classification using CT images alone, (2) classification using CT images and early phase PET images, and (3) classification using CT images and early and delayed phase PET images.

Results

To confirm whether individual feature values were useful for distinguishing between benign and malignant nodules, we calculated the mean, median, and standard deviation (SD) of the values in the two groups. Then, effect sizes [33] were calculated. Furthermore, t-values and p-values were calculated using the t-test (double-sided test). The results are listed in Table 2.

There was a significant difference between the characteristic features of benign and malignant nodules with regard to SUV. The SUV obtained by PET examination was the most significant and was confirmed as useful for distinguishing between benign and malignant cells. In addition, some texture features showed significant differences between benign and malignant nodules, indicating their effectiveness. The differences in features with regard to spicules were of low significance. Furthermore, the average CT_{max} in benign nodules was higher than that in malignant nodules.

Then, the nodule classification scheme was evaluated using the ROC curve. In the curve, the TP rate was defined as the ratio of the number of corrected malignancy nodules to the total number of malignancy nodules. FP rate was defined as the ratio of the number of misclassified benign nodules to the total number of benign nodules. Furthermore, performance was evaluated by the leave-one-out cross-validation method.

To analyze the distinguishing characteristics of our method, we evaluated three methods: (1) classification using CT images alone, (2) classification using CT images and early phase PET images, and (3) classification using CT images and early and delayed phase PET images. The ROC curves for each of the above methods are shown in Fig. 11. The area under the curve (AUC) for methods (1), (2), and (3) was 0.730, 0.860, and 0.895, respectively. Considering the accuracy rate of malignant nodules being 0.944, the accuracy rates of benign nodules for methods (1), (2), and (3) were 0.277, 0.611, and 0.722, respectively. The CT and PET images in the trans-axial plane (with accuracy rates for benign and malignant nodules of 0.722 and 0.944, respectively) are shown in Fig. 10. Regarding the significant differences among the three ROC curves, the p-values between (1) and (2), (2) and (3), and (1) and (3) were 0.032, 0.104, and 0.021, respectively.

Discussion

As can be seen in Table 2, there was a significant difference between the characteristic features of the benign and malignant nodules with regard to SUV. The SUV obtained by PET examination was confirmed to be useful for distinguishing between benign and malignant. However, inflammatory diseases can also produce high SUVs; hence, SUVs should be combined with other features for correct classification. In addition, some texture features were effective for distinguishing between benign and malignant nodules. The difference in features with regard to spicules was of low significance. This is because many spicules are observed even in inflammatory diseases. Furthermore, the average CT_{max} in benign nodules was higher than that in malignant nodules. This is due to calcification inside the benign nodules. The distributions of all characteristic features over-

Table 2 Basic statistics and t-test results

Feature	Benign			Malignant			t-value	p-value	Effect size
	Mean	Median	SD	Mean	Median	SD			
$ESUV_{max}$	3.31	2.28	2.27	11.66	9.95	6.58	5.242	<0.001	3.685
$DSUV_{max}$	3.81	2.37	3.08	14.29	11.76	8.3	5.165	<0.001	3.395
$ESUV_{peak}$	2.31	1.69	1.42	7.66	5.49	4.78	4.684	<0.001	3.764
$DSUV_{peak}$	2.51	1.56	1.79	9.46	6.83	6.29	4.638	<0.001	3.886
ΔSUV_{max}	0.5	0.21	0.88	2.63	1.96	1.87	4.49	<0.001	2.4
ΔSUV_{peak}	0.2	0.05	0.44	1.8	1.05	2.04	3.348	0.003	3.663
T_{4_0}	0.0651	0.0602	0.0239	0.045	0.0418	0.0117	3.289	0.003	0.839
T_{4_90}	0.0655	0.0614	0.0234	0.0459	0.0437	0.0114	3.274	0.003	0.834
T_{5_0}	0.0345	0.0305	0.0192	0.0209	0.0184	0.0084	2.837	0.009	0.708
D_{xy}	17.2	17.5	3.8	22.1	22.8	6.7	2.771	0.01	1.279
T_{5_90}	0.0342	0.0301	0.0179	0.0211	0.0193	0.008	2.92	0.08	0.732
C_b	0.0811	0.0844	0.0305	0.0941	0.0964	0.0137	1.692	0.104	0.425
T_{2_0}	33.53	33.9	8.27	36.8	37.21	3.09	1.615	0.121	0.394
L_{mass}	56	22.6	104.4	193.2	47.1	365.9	1.574	0.131	1.313
CT_{max}	983.9	661.3	710.7	727.3	535.5	448.8	1.333	0.193	0.361
T_{3_0}	0.63	0.668	0.154	0.575	0.576	0.104	1.31	0.2	0.362
T_{2_90}	34.36	32.58	8.63	37.02	35.3	4.408	1.198	0.242	0.308
T_{1_0}	2616.2	2536.8	1096	2885.4	2855.9	395.9	1.009	0.324	0.245
D_z	18.1	16.5	7.9	20.6	19.5	8.1	0.955	0.346	0.312
CT_{centre}	0.3	19.9	151	26.1	35.9	34.1	0.729	0.475	0.171
T_{3_90}	0.612	0.674	0.187	0.58	0.604	0.093	0.671	0.508	0.171
T_{1_90}	2734.4	2500.2	1205.3	2916.3	2761.6	619.5	0.586	0.563	0.15
SP_2	0.0195	0.0216	0.0104	0.0147	0.0184	0.0161	0.359	0.722	0.101
SP_1	73.7	42	103.5	82.7	75.5	52.8	0.338	0.738	0.086
L_{line}	42.5	24.4	57.8	43.4	28.9	33.6	0.061	0.952	0.016

$ESUV_{max}$ maximum standardized uptake value (SUV) of early PET images, $DSUV_{max}$ maximum SUV of delayed PET images, ΔSUV_{max} difference between $ESUV_{max}$ and $DSUV_{max}$, $ESUV_{peak}$ peak SUV of early PET images, $DSUV_{peak}$ peak SUV of delayed PET images, ΔSUV_{peak} difference between $ESUV_{peak}$ and $DSUV_{peak}$, CT_{max} maximum CT value inside the nodule, CT_{centre} CT value at the center of the nodule, D_{xy} and D_z nodule diameters in trans-axial plane and z-directions, respectively, L_{mass} ball-like feature, L_{line} line-like feature, $C_b CT_{R1}$-CT_{R2}, SP_1 and SP_2 features of spicules, T_{1_0}~T_{5_0} texture features for horizontal direction, T_{1_90}~T_{5_90} texture features for vertical direction

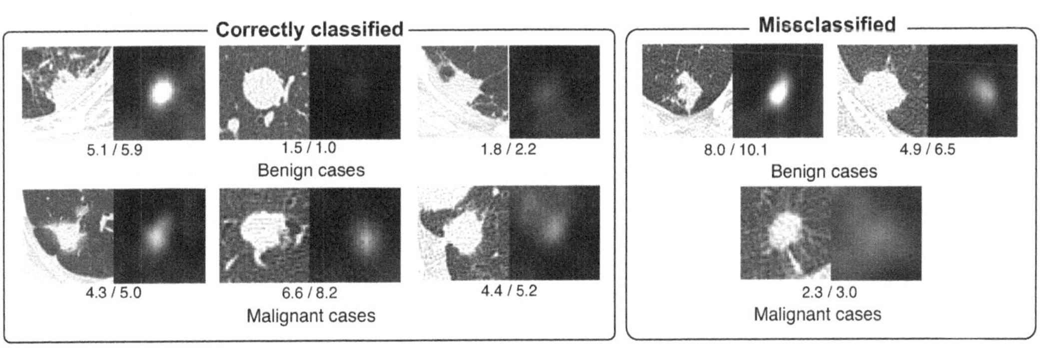

Fig. 10 Classification results. Images on the left are CT images, and images on the right are early phase PET images. The two values under each image are the standardized uptake values (SUVs) of the early and delayed PET images (early/delayed)

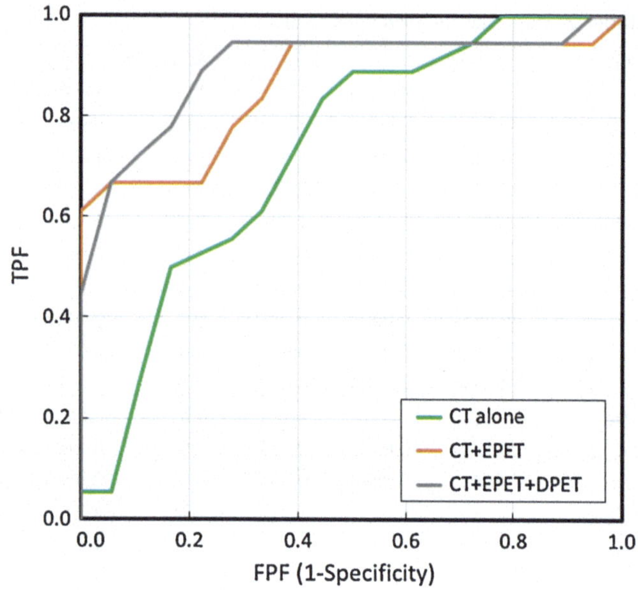

Fig. 11 ROC curves of classification

lapped between the two classes, which makes it difficult to distinguish between the two classes using a single feature. Therefore, integration of multiple characteristic features using a classifier would be more effective.

From the ROC curves (Fig. 11), the proposed method using the CT image and two-phase PET images shows that the incorrect rate of detection for benign nodules was 0.278 (accuracy rate of 0.722 for benign nodules), whereas the accuracy rate of malignant nodule detection was 0.944. The target nodules in this study were "difficult nodules" for differentiation by image diagnosis using CT and PET/CT. Most of the benign cases were not confirmed in follow-up examinations but were confirmed after biopsy. Our results indicate that biopsy examination with its accompanying physical hardship to the patient, especially in benign cases, could be reduced by 72.2%.

In addition, compared to classification using only CT images, the result of using PET images together with CT improved the AUC of the ROC curves, as shown in Fig. 11. This proves the effectiveness of using both anatomical and functional information together. Based on the results of the *p*-values of the ROC curves, we found a significant difference between analysis using CT alone and analysis using both CT and PET images.

These results indicate that the proposed method may be useful for improving the accuracy of malignancy analysis.

Automated Malignancy Analysis Using Lung Cytological Images

Introduction

To improve the survival rate among lung cancer patients, early-stage detection and treatment are necessary. CT examination is an effective diagnostic tool for the early detection of lung cancer [1, 2]. If a suspicious lesion is found by CT examination, a subsequent pathological diagnosis is performed to check the lesion in detail.

To give a pathological diagnosis, cytological diagnosis is first conducted using sputum, pleural effusion, and brush during bronchoscopic biopsy [34]. However, this is performed by cytotechnologists and cytopathologists in a procedure that presents some challenges. First, these experts must detect abnormal cells among many cells, and variations in cell morphology further complicate the detection. The final diagnosis by the cytopathologist usually confirms specimens with a suspected abnormality pointed out by the cytotechnologist. Therefore, FNs should be minimized at the first stage. Although cytopathologists play an important role in precision medicine, there is global shortage of such professionals [35], which is a burden on early diagnosis. Therefore, the automatic analysis of cytological images to quantify the grade of malignancy from samples may improve diagnostic accuracy and shorten diagnosis time.

In this study, we developed an automated method for classification of lung cells in cytological images. The automated classification for benign and malignant cells in cytology is an important computer assisted technology not only for cytopathologists but also for screeners. However, to the best of our knowledge, the classification of benign and malignant lung cells using deep learning has not been implemented to date. Therefore, in this section, we propose an automated method for classification for benign and malignant cells in lung cytological images [36].

Materials and Methods

Image Dataset

We collected 46 cases of lung tissues and cells by interventional cytology under either bronchoscopy or CT-guided fine needle aspiration cytology at the Fujita Health University Hospital (Toyoake, Japan) from 2012 to 2014, comprising 18 benign and 28 malignant cases. The cytological specimens were prepared with liquid-based cytology using the BD SurePathTM Liquid-based Pap Test (Beckton Dickinson and Company, Franklin Lakes, New Jersey) and they were stained by the Papanicolaou method. Using a digital still camera with matrix size of 1280 × 960 pixels (DP20, Olympus Corporation, Tokyo, Japan) attached to a microscope (BX53, Olympus Corporation) with ×40 objective lens, we collected 197 images of benign and 220 images of malignant cells. To generate the image dataset for the DCNN, patch images of 224 × 224 pixels were cut from the original microscopic images without overlapping. The patch images were checked by a cytopathologist; an image with at least one malignant cell was considered to be malignant. Consequently, only when all cells were benign was the whole image considered as benign. The final dataset consisted of 621 patch images, 306 of benign and 315 of malignant cells.

Network Architecture

We employed the VGG-16 DCNN model for classification using patch images. It was released in 2014 by the Visual Geometry Group at the University of Oxford [37]. This model achieved second place in the 2014 ImageNet Large-Scale Visual Recognition Challenge due to its high performance, despite it being a sequential model with very simple architecture. In this study, we fine-tuned a VGG-16 model pretrained on the ImageNet dataset, which has a much larger number of training image samples than our original dataset. For classification, we replaced the fully connected layers of the original VGG 16 model with three layers having 1024, 256, and 2 units, as shown in Fig. 12. The output was given by a

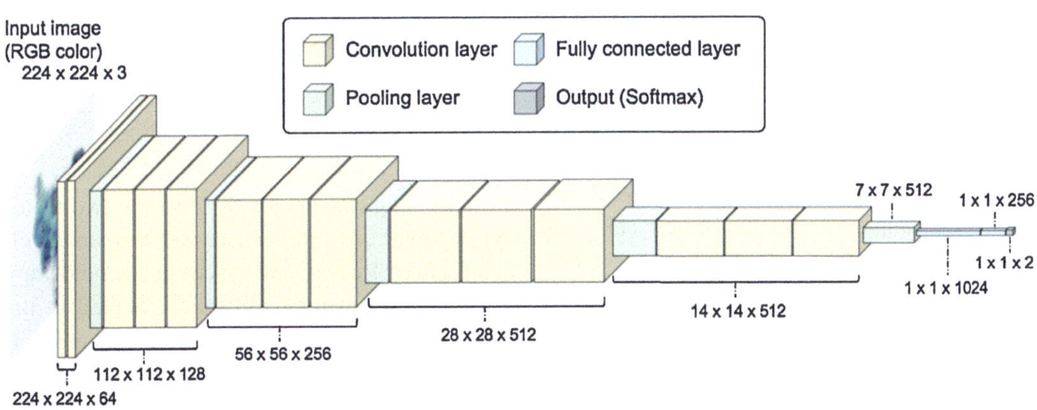

Fig. 12 DCNN architecture for classification of lung cytological images (VGG-16 model)

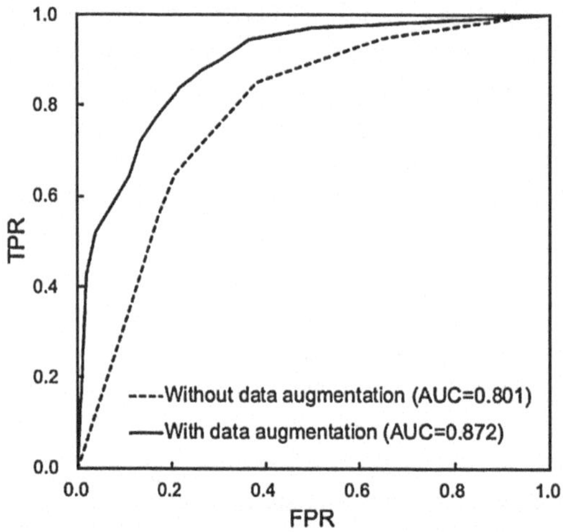

Fig. 13 ROC curve of patch-based classification

softmax layer. The parameters of the replaced layers were randomly initialized and fitted without changing the parameters of other layers.

The proposed method was implemented using the Keras and TensorFlow artificial intelligence platforms. Training was conducted using a mini-batch size of 32, optimization based on stochastic gradient descent (SGD), a learning rate of 10^{-5}, and a momentum of 0.9. We set batch size as 32 out of 16, 32, 64, and 128 so that training loss smoothly decreased.

Results and Discussion

The ROC curves of the patch-based classification are shown in Fig. 13. Also, Table 3 describes the confusion matrix of the patch-based classification. The AUCs of the patch-based classification using the original and augmented images were 0.801 and 0.872, respectively. The accuracies of the patch-based classification for steps one, two, and three of a threefold cross validation were 86.9%, 81.0%, and 70.0%, respectively. Figure 14 shows sample images of benign cells and malignant cells.

As a result of experiments, classification accuracy regarding the analyzed cases was approximately 80.0%. Compared to other CNN architectures, VGG-16 has the highest classification ability. Thus, the proposed method provides complementary characteristics to diagnosis by cytopathologists and retrieves high malignancy discrimination rates.

Table 3 Confusion matrix of patch-based classification

		Predicted		Accuracy [%]	
		Benign	Malignant		
Actual	Benign	194	112	63.4	Overall: 79.2
	Malignant	17	298	94.6	

In future work, to improve classification accuracy, we will increase the number of cases for training the DCNN. Furthermore, it is possible to apply a DCNN with more layers. In this study, we acquired images using a microscope with a ×40 objective lens and a standard-resolution camera for general pathological examination. However, improved accuracy can be expected by classifying images obtained from a high-magnification objective lens and a high-resolution camera. In addition, in the practical diagnosis of cytology, it is difficult to clearly classify cells as benign or malignant due to atypical cells. In the future, we would like to consider adding the category of "atypical cells" to the actual diagnosis.

In this section, we proposed an automated DCNN-based classification scheme for malignant lung cells using microscopic images. Evaluation results retrieved a sensitivity, specificity, and accuracy of 94.6%, 63.4%, and 79.2%, respectively. These results reveal that the proposed method can effectively identify malignant cells in microscopic cytological images.

Automated Classification of Lung Cancer Types from Cytological Images

Introduction

Primary lung cancers are divided into two major types: small cell lung cancer and non-small cell lung cancer. Recently, advances in chemotherapy and radiation therapy [38] have resulted in the latter being further classified into adenocarcinoma, squamous cell carcinoma, and large cell carcinoma [39]. It is often difficult to exactly

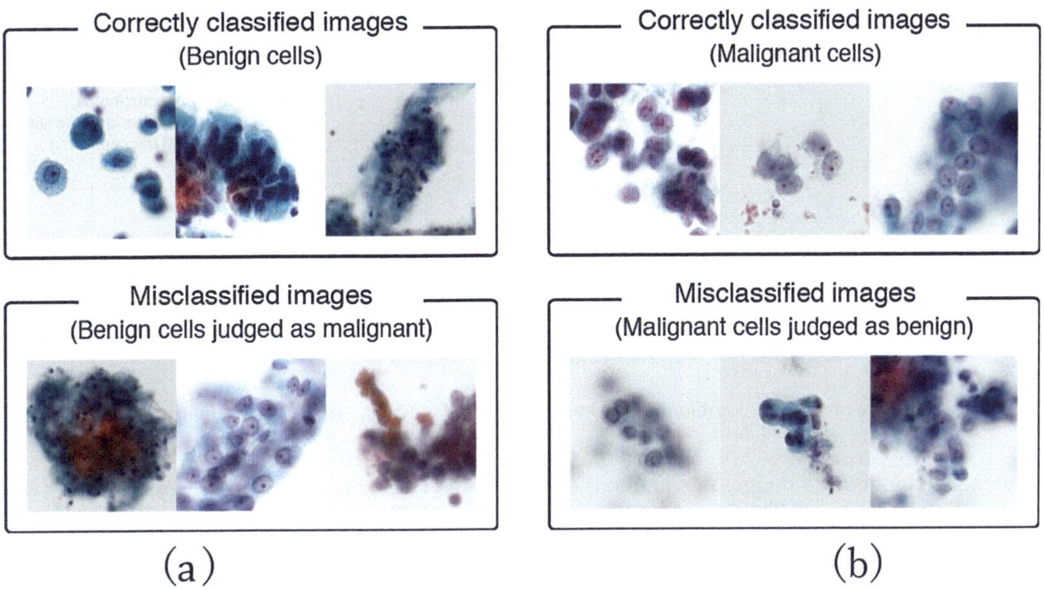

Fig. 14 Sample images of (**a**) benign cells and (**b**) malignant cells

differentiate between adenocarcinoma and squamous cell carcinoma in terms of morphological characteristics, meaning an immunohistochemical evaluation is required. Furthermore, there are many varieties of morphologies among these cancer cells. Therefore, computer-aided diagnosis (CAD) can be a useful tool for avoiding misclassification. Among the four major types of lung cancer, large cell carcinoma is the easiest to detect because it has severe atypism. In this study, we focused on automated classification of cancer types (adenocarcinoma, squamous cell carcinoma, and small cell carcinoma) using microscopic images for cytology.

Various studies that apply CAD methods to pathological images have been conducted [40–43]. However, to the best of our knowledge, no method has been developed to classify lung cancer types from cytological images. Also, "deep learning" is well known to give better performances than conventional image classification techniques. For example, Krizhevsky et al. [13] won the 2012 ImageNet Large-Scale Visual Recognition Challenge using a DCNN to classify high-resolution images. Many research groups have investigated the application of DCNNs to medical images [11, 44–46]. In this study, we developed an automated classification scheme for lung cancers in microscopic images using a DCNN [47].

Materials and Methods

Image Dataset

76 cases of cancer cells were collected by exfoliative or interventional cytology under bronchoscopy or CT-guided fine needle aspiration cytology. They consisted of 40 cases of adenocarcinoma, 20 cases of squamous cell carcinoma, and 16 cases of small cell carcinoma. Final diagnosis was made in all cases via a combination of histopathological and immunohistochemical diagnoses.

The cytological specimens were prepared with a liquid-based cytology (LBC) system using the BD SurePath liquid-based Pap Test (Beckton Dickinson, Durham, NC, USA) and were stained using the Papanicolaou method. Using a digital still camera (DP70, Olympus, Tokyo, Japan) attached to a microscope (BX51, Olympus) with a ×40 objective lens, 82 images

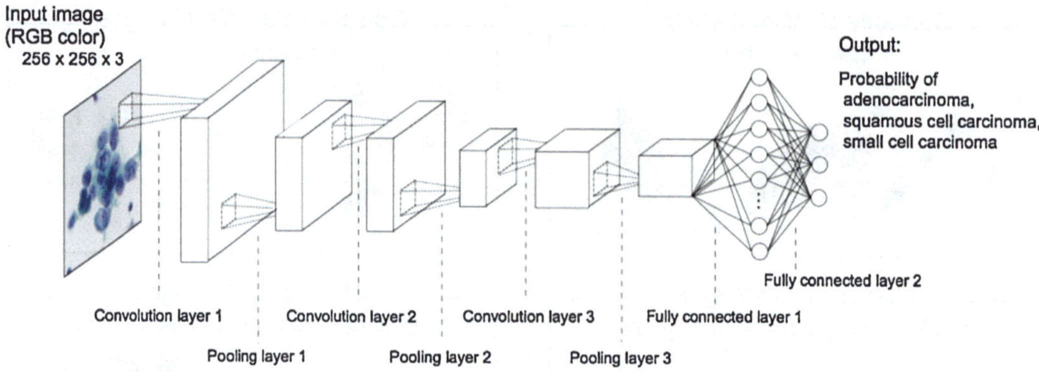

Fig. 15 Architecture of the deep convolutional neural network used for cancer-type classification

of adenocarcinoma, 125 images of squamous cell carcinoma, and 91 images of small cell carcinoma were collected in JPEG format. The initial matrix size of each JPEG image was 2040 × 1536 pixels.

Subsequently, 768 × 768-pixel square images were generated by cropping. Finally, they were resized to 256 × 256 pixels.

Network Architecture

The architecture of the DCNN used for cancer-type classification is illustrated in Fig. 15. It consists of three convolution layers, three pooling layers, and two fully connected layers. Color microscopic images are fed to the input layer of the DCNN. The filter size, number of filters, and stride for each layer are specified in Table 4. In the last layer, the probabilities of cancer types (adenocarcinoma, squamous cell carcinoma, and small cell carcinoma) are obtained using a softmax function. During training, we employed the dropout method (dropout rate = 50% for fully connected layers) to prevent overfitting.

Table 4 Filter size, number of filters, and stride for each layer of DCNN

Layer	Kernel	Stride	Output size
Input	–	–	256 × 256
Convolution 1	5 × 5 × 3	1 × 1	256 × 256 × 32
Pooling 1	3 × 3	2 × 2	128 × 128 × 32
Convolution 2	5 × 5 × 32	1 × 1	128 × 128 × 32
Pooling 2	3 × 3	2 × 2	64 × 64 × 32
Convolution 3	5 × 5 × 32	1 × 1	64 × 64 × 64
Pooling 3	3 × 3	2 × 2	32 × 32 × 64
Fully connected 1	–	–	64
Fully connected 2	–	–	3

Results and Discussion

For classification of the three cancer types, the DCNN was trained and evaluated using augmented data. Its classification performance was evaluated via threefold cross validation. In this process, 298 images were randomly divided into three groups. However, images taken from the same specimen belonged to the same group. By data augmentation, the number of images in each class was unified to approximately 5000. Figure 16 shows sample images of correctly classified and misclassified cancer types obtained using the proposed method. The classification confusion matrix is described in Table 5. It can be seen that squamous cell cancer was often mistaken for adenocarcinoma.

As a result of experiments, 70% of lung cancer cells were classified correctly. Most of the correctly classified images had typical cell morphologies and arrangements. In traditional cytology, pathologists perform classification of small cell carcinoma and non-small cell carcinoma. The classification accuracy rate in this case is 255/298 = 85.6%, which is considered to be sufficient. Of the three types of lung cancers, classification accuracy was highest for adenocarcinoma and lowest for squamous cell carcinoma. This result may be related to the variation of images used for training. Therefore,

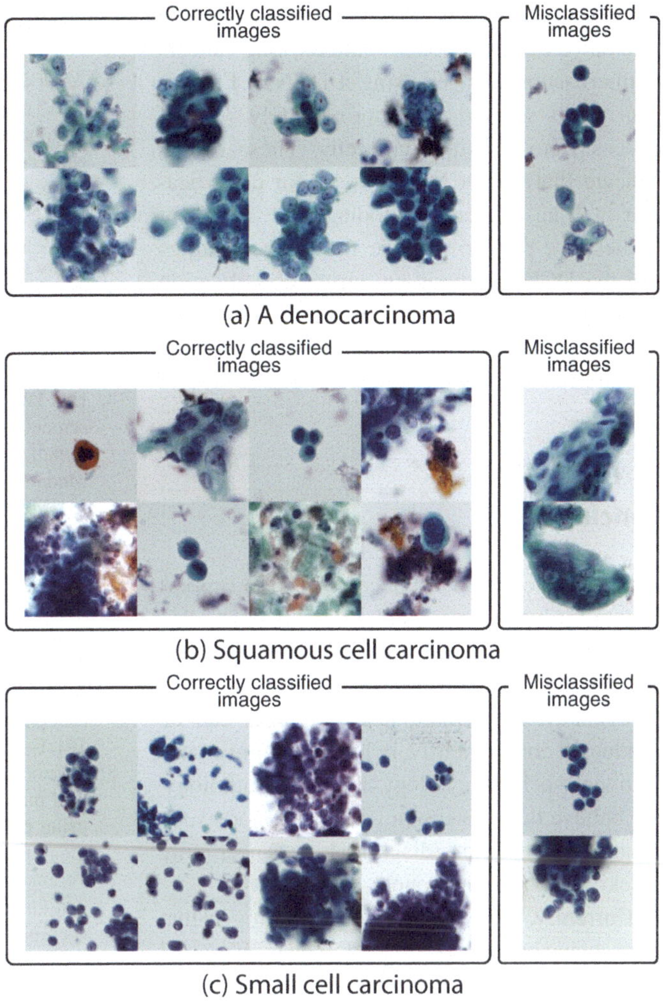

Fig. 16 Sample images of correctly classified and misclassified carcinomas. (**a**) adenocarcinoma; (**b**) squamous cell carcinoma; (**c**) small cell carcinoma

Table 5 Confusion matrix of classification results

	Adenocarcinoma	Squamous cell carcinoma	Small cell carcinoma
Adenocarcinoma	73 (89.0%)	8 (9.8%)	1 (1.2%)
Squamous cell carcinoma	35 (28.0%)	75 (60.0%)	15 (12.0%)
Small cell carcinoma	7 (7.7%)	20 (22.0%)	64 (70.3%)

we plan to increase the number of cases of adenocarcinoma and squamous cell carcinoma used in future studies. Small cell carcinoma has distinctive features, such as only a few cytoplasms and a small nucleus. Therefore, its image characteristics could be understood by the DCNN from even a small number of images.

The classification accuracy of 71.1% is comparable to that of a cytotechnologist or pathologist [48, 49]. It is worth noting that the DCNN is able to understand cell morphology and placement of cancer cells solely from images without prior knowledge or experience of biology and pathology. This is a satisfactory result because cytological diagnosis of lung cancer is a difficult task for pathologists. Therefore, our method will be useful in assisting cytological examinations for lung cancer diagnosis.

To summarize this section, we developed an automated classification scheme for lung cancers in microscopic images using a DCNN. Evaluation results showed that approximately 70% of images were classified correctly. These results indicate that the DCNN is useful for classification of lung cancer in cytodiagnosis. In future studies, we plan to analyze the image features that the DCNN focuses on during classification in order to reveal the classification mechanism in detail. Also, we hope to develop a method to comprehensively classify cells and arrays of cells.

Conclusion

In this chapter, we introduced a decision support system for lung cancer using PET/CT and microscopic images. PET/CT gives useful information for detection of lung lesions and evaluation of malignancy, and pathological diagnosis can be conducted comprehensively by combining examinations, such as cytology and histopathology. We believe that this system, which supports such diagnoses, will help to diagnose lung cancer more quickly and more accurately.

Currently, we are developing technologies that can achieve better performance, even with a small amount of data [50]. Because each of the introduced methods was evaluated using different cases, in the future we will proceed with verification of whether the pipeline from detection to definitive diagnosis works effectively using many cases.

Acknowledgements The authors are grateful for the help of Dr. Tsuneo Tamaki, Dr. Masami Nishio, Mr. Osamu Yamamuro, Mr. Katsuaki Takahashi, and Mr. Toshiki Kobayashi of the Nagoya Radiological Diagnosis Foundation; Dr. Yuka Kiriyama and Dr. Takahiro Inoue, and Mr. Masakazu Tsujimoto from Fujita Health University Hospital; Dr. Ronald M. Summers and Dr. Ke Yan from the National Institute of Health; and Dr. Ling Zhang from NVIDIA corporation.

This research was supported in part by a Grant-in-Aid for Scientific Research on Innovative Areas (No. 26108005) and a Grant-in-Aid for Scientific Research (No. 17K09070), MEXT, Japan.

References

1. American Cancer Society, Cancer Facts and Figures (2015). https://www.cancer.org/content/dam/cancer-org/research/cancer-facts-and-statistics/annual-cancer-facts-and-figures/2015/cancer-facts-and-figures-2015.pdf Accessed 2 July 2019
2. Sone S, Takashima S, Li F et al (1998) Mass screening for lung cancer with mobile spiral computed tomography scanner. Lancet 351:1242–1245
3. The National Lung Screening Trial Research Team (2011) Reduced lung-cancer mortality with low-dose computed tomographic screening. N Engl J Med 365:395–409
4. Lee JW, Kang KW, Paeng JC et al (2009) Cancer screening using 18F-FDG PET/CT in Korean asymptomatic volunteers: a preliminary report. Ann Nucl Med 23:685–691
5. Ide M, Suzuki M (2005) Is whole-body FDG-PET valuable for health screening? Eur J Nucl Med Mol Imaging 32:339–341
6. Wever W, Meylaerts L, Ceuninck L et al (2007) Additional value of integrated PET-CT in the detection and characterization of lung metastases: correlation with CT alone and PET alone. Eur Radiol 17:467–473
7. Johnston WW (1986) Cytologic diagnosis of lung cancer: principles and problems. Pathol Res Pract 181:1–36
8. Teramoto A, Fujita H, Takahashi K et al (2014) Hybrid method for the detection of pulmonary nodules using positron emission tomography / computed tomography: a preliminary study. Int J CARS 9:59–69
9. Teramoto A, Adachi H, Tsujimoto M et al (2015) Automated detection of lung tumors in PET/CT images using active contour filter. In: Proceedings of SPIE medical imaging 2015: computer-aided diagnosis 9414:94142V-1 - 94142V-6
10. Teramoto A, Fujita H (2018) Automated lung nodule detection using positron emission tomography/-computed tomography. In: Suzuki K, Chen Y (eds) Artificial intelligence in decision support systems for diagnosis in medical imaging. Intelligent systems reference library, vol 140. Springer, Cham
11. Teramoto A, Fujita H, Yamamuro O et al (2016) Automated detection of pulmonary nodules in PET/CT images: ensemble false-positive reduction using a convolutional neural network technique. Med Phys 43(6):2821–2827
12. LeCun Y, Bengio Y, Hinton GE (2015) Deep learning. Nature 521:436–444
13. Krizhevsky A, Sutskever I, Hinton GE (2012) ImageNET classification with deep convolutional neural networks. Adv Neur Inf 25:1106–1114
14. He K, Zhang X, Ren S et al (2016) Deep residual learning for image recognition. In: 2016 IEEE conference on computer vision and pattern recognition (CVPR), Las Vegas, NV, 2016, pp 770–778

15. Keyes JW (1995) SUV: standard uptake or silly useless value? J Nucl Med 36(10):1836–1839
16. Chang CC, Lin CJ. LIBSVM: A library for support vector machines. Software available at http://www.csie.ntu.edu.tw/cjlin/libsvm/
17. Jia Y, Shelhamer E, Donahue J, et al (2014) Caffe: convolutional architecture for fast feature embedding. In: ACM conference on multimedia, pp 675–678
18. Gould MK, Maclean CC, Kuschner WG et al (2001) Accuracy of positron emission tomography for diagnosis of pulmonary nodules and mass lesions: a meta-analysis. JAMA 285(7):914–924
19. Armato SG 3rd, Altman MB, Wilkie J et al (2003) Automated lung nodule classification following automated nodule detection on CT: a serial approach. Med Phys 30(6):1188–1197
20. Way TW, Hadjiiski LM, Sahiner B et al (2006) Computer-aided diagnosis of pulmonary nodules on CT scans: segmentation and classification using 3D active contours. Med Phys 33(7):2323–2337
21. Zhang F, Song Y, Cai W et al (2014) Lung nodule classification with multilevel patch-based context analysis. IEEE Trans Biomed Eng 61(4): 1155–1166
22. Madero Orozco H, Vergara Villegas OO, Cruz Sánchez VG et al (2015) Automated system for lung nodules classification based on wavelet feature descriptor and support vector machine. Biomed Eng Online 14:9
23. Shen W, Zhou M, Yang F et al (2017) Multi-crop convolutional neural networks for lung nodule malignancy suspiciousness classification. Pattern Recogn 61:663–673
24. Nie Y, Li Q, Li F et al (2006) Integrating PET and CT information to improve diagnostic accuracy for lung nodules: a semiautomatic computer-aided method. J Nucl Med 47(7):1075–1080
25. Teramoto A, Tsujimoto M, Inoue T et al (2018) Automated classification of pulmonary nodules through a retrospective analysis of conventional CT and two-phase PET images in patients undergoing biopsy. Asia Oceania J Nucl Med Biol 7(1):29–37
26. MacMahon H, Naidich DP, Goo JM et al (2017) Guidelines for management of incidental pulmonary nodules detected on CT images: from the Fleischner society 2017. Radiology 284(1):161659
27. Sim YT, Poon FW (2013) Imaging of solitary pulmonary nodule—a clinical review. Quant Imaging Med Surg 3(6):316–326
28. Li Q, Sone S, Doi K (2003) Selective enhancement filters for nodules, vessels, and airway walls in two- and three-dimensional CT scans. Med Phys 30(8):2040–2051
29. Rangayyan RM, Ayres FJ (2006) Gabor filter and phase portraits for the detection of architectural distortion in mammograms. Med Biol Eng Comput 44(10):883–894
30. Yoshikawa R, Teramoto A, Matsubara T et al (2014) Automated detection of architectural distortion using improved adaptive Gabor filter. Breast Imaging 8539:606–611
31. Haralick RM, Shanmugam K, Dinstein I (1973) Textural features for image classification. IEEE Trans Syst Man Cybern SMC-3(6):610–621
32. Breiman L (2001) Random forests. Mach Learn 45(1):5–32
33. Cohen J (1988) Statistical power analysis for the behavioral sciences, 2nd edn. Lawrence Erlbaum Associates, Hillsdale
34. Johnston WW (1986) Cytologic diagnosis of lung cancer: principles and problems. Pathol Re Pract 181:1–36
35. Robboy SJ, Weintraub S, Horvath AE et al (2013) Pathologist workforce in the United States: I. development of a predictive model to examine factors influencing supply. Arch Pathol Lab Med 137:1723–1732
36. Teramoto A, Yamada A, Kiriyama Y et al (2019) Automated classification of benign and malignant cells from lung cytological images using deep convolutional neural network. Inf Med Unlocked 16:100205
37. Simonyan K, Zisserman A (2015) Very deep convolutional networks for large-scale image recognition. arXiv:1409.1556
38. Baas P, Belderbos JS, Senan S et al (2006) Concurrent chemotherapy (carboplatin, paclitaxel, etoposide) and involved-field radiotherapy in limited stage small cell lung cancer: a Dutch multicenter phase II study. Br J Cancer 94(5):625–630
39. Travis WD, Brambilla E, Burke A et al (2015) WHO classification of tumours of the lung, pleura, thymus, and heart, 4th edn. IARC, Lyon
40. He L, Long LR, Antani S et al (2012) Histology image analysis for carcinoma detection and grading. Comput Methods Prog Biomed 107(3): 538–556
41. Barker J, Hoogi A, Depeursinge A et al (2016) Automated classification of brain tumor type in whole-slide digital pathology images using local representative tiles. Med Image Anal 30:60–71
42. Ojansivu V, Linder N, Rahtu E et al (2013) Automated classification of breast cancer morphology in histopathological images. Diagn Pathol 8(1):S29
43. Ficsor L, Varga VS, Tagscherer A et al (2008) Automated classification of inflammation in colon histological sections based on digital microscopy and advanced image analysis. Cytometry 73A(3): 230–237
44. Zhang L, Le L, Nogues I et al (2917) DeepPap: deep convolutional networks for cervical cell classification. IEEE J Biomed Health Inform 21(6):1633–1643
45. Cha KH, Hadjiiski L, Samala RK et al (2016) Urinary bladder segmentation in CT urography using deep-learning convolutional neural network and level sets. Med Phys 43:1882–1896

46. Yan K, Wang X, Lu L et al (2018) DeepLesion: automated mining of large-scale lesion annotations and universal lesion detection with deep learning. J Med Imaging 5(3):036501
47. Teramoto A, Tsukamoto T, Kiriyama Y et al (2017) Automated classification of lung cancer types from cytological images using deep convolutional neural networks. Biomed Res Int 2017(4067832): 1–6
48. Nizzoli R, Tiseo M, Gelsomino F et al (2011) Accuracy of fineneedle aspiration cytology in the pathological typing of non-small cell lung cancer. J Thorac Oncol 6(3):489–493
49. Sigel CS, Moreira AL, Travis WD et al (2011) Subtyping of non-small cell lung carcinoma: a comparison of small biopsy and cytology specimens. J Thorac Oncol 6(11):1849–1856
50. Onishi Y, Teramoto A, Tsujimoto M et al (2019) Automated pulmonary nodule classification in computed tomography images using a deep convolutional neural network trained by generative adversarial networks. BioMed Res Int 2019(6051939):1–9

Lesion Image Synthesis Using DCGANs for Metastatic Liver Cancer Detection

Keisuke Doman, Takaaki Konishi, and Yoshito Mekada

Abstract

This chapter proposes a method to detect metastatic liver cancer from X-ray CT images using a convolutional neural network (CNN). The proposed method generates various lesion images by the combination of three kinds of generation methods: (1) synthesis using Poisson Blending, (2) generation based on CT value distributions, and (3) generation using deep convolutional generative adversarial networks (DCGANs). The proposed method constructs two kinds of detectors by using synthetic (fake) lesion images generated by the methods as well as real ones. One of the detectors is a 2D CNN for detecting candidate regions in a CT image, and the other is a 3D CNN for validating the candidate regions. Experimental results showed that the proposed method gave 0.30 improvement from 0.65 to 0.95 in terms of the detection rate, and 0.70 improvement from 0.90 to 0.20 in terms of the number of false detections per case. From the results, we confirmed the effectiveness of the proposed method.

Keywords

Cancer diagnosis · Metastatic liver cancer · Cancer detection · CT image · Lesion image synthesis · CNN · DCGAN · Poisson Blending

Introduction

Machine learning-based lesion detection is useful for liver cancer diagnosis, because cancer diagnosis is mainly performed by doctor's visual check and takes much time. In the field of object recognition, highly accurate models of convolutional neural networks (CNNs), such as VGG [1], inception [2], and ResNet [3], were proposed. We focus on a CNN-based method to accurately detect liver cancer lesions. These machine learning techniques including CNNs, however, generally require a massive number of training images, and it is difficult to collect such CT or MRI case images.

Data augmentation is performed to increase the appearance variation of training images by applying geometric transformations such as translation and rotation. A lesion actually has an arbitrary size and occurs at an arbitrary position in an organ, and moreover the internal condition

K. Doman (✉) · Y. Mekada
Graduate School of Engineering, Chukyo University, Toyota, Japan
e-mail: kdoman@sist.chukyo-u.ac.jp; y-mekada@sist.chukyo-u.ac.jp

T. Konishi
Graduate School of Computer and Cognitive Sciences, Chukyo University, Toyota, Japan

Persol Research & Development Co., Ltd., Tokyo, Japan

G. Lee, H. Fujita (eds.), *Deep Learning in Medical Image Analysis*, Advances in Experimental Medicine and Biology 1213, https://doi.org/10.1007/978-3-030-33128-3_6

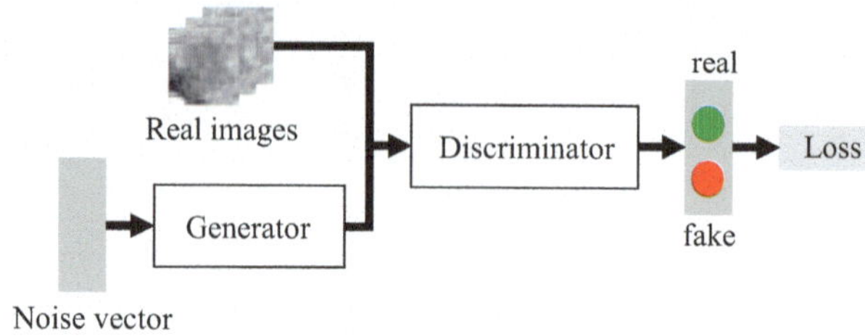

Fig. 1 GANs architecture

of a lesion differs in individual cases. Therefore, simple data augmentation cannot cover a variety of the lesion appearance.

In this chapter, we propose the combination of three kinds of methods for generating pseudo-lesions at arbitrary position and with an arbitrary size, and show that it is effective to use a large amount of images generated by the proposed method for metastatic liver cancer detection.

One of the generation methods uses Poisson Blending [4] in order to overlay lesion images with real lesion areas on healthy subjects' CT images [5]. Although the method can improve the detection accuracy, it cannot generate various appearances of liver cancers because it does not add any change to overlaid lesions. Another one of the generation methods generates lesion images considering the CT value distribution of a liver cancer region [6]. Most of small lesions have spherical shapes, and their concentration profiles often have bathtub-type distributions of CT values. The method generates lesion images based on the distribution, and consequently, can improve the detection accuracy. However, the problem on over-fitting of a resultant detector may be caused by using images generated by the method because of the lack of variation in the generated lesions. The last one of the generation method uses generative adversarial networks (GANs) [7, 8]. As shown in Fig. 1, GANs include two kinds of networks: one is a generator that generates fake but real-looking images, and the other is a discriminator that distinguishes real images from fake ones generated by the generator. Although the generator is designed to generate images similar to real ones in cooperation with the discriminator, GANs have a problem on the convergence of the learning process, which leads to the difficulty of generating high-quality images. A model of deep convolutional generative adversarial networks (DCGANs) [9] with convolution layers was proposed as one of the solutions. In the framework of DCGANS, noise vectors are input to a generator network, and fake images are generated by using a CNN with up-sampling layers. On the other hand, a discriminator network is a CNN that distinguished between real and fake images. In the same mechanism as GAN, the generator network can finally generate real-looking images, whereas the discriminator network can finally recognize fake images as fake.

As for metastatic liver cancer detection from X-ray CT images, a method using fully convolutional networks (FCN) was proposed [10]. The method, however, has a problem on the detection accuracy for small lesions with the low liver-lesion contrast. For accurate metastatic liver cancer detection, it is necessary to construct a detector with training images containing the 3D shape features of liver cancer, which requires more variety of lesion images than the case of 2D liver cancer detection. Regarding this issue, we propose a two-stage method for detecting metastatic liver cancer: the first stage is to detect candidate regions using a 2D CNN, and the second stage is to validate the candidate regions using a 3D CNN. We report the effectiveness of the proposed method for 20 cases of liver cancer with lesions of equal to or less than 20 mm in diameter.

This chapter is organized as follows. Section "Proposed Method" describes the proposed method. Next, experimental results are reported in section "Experiments." Finally, section "Conclusion" concludes this chapter.

Proposed Method

The proposed method is composed of two steps: one is a training step, and the other is a detection step. Each step uses abdominal CT images of the portal phase as shown in Fig. 2. The dataset used in this research and the details of each step are described below.

Dataset

We use the abdominal CT images of the portal phase, released through the CAD contest of the Japanese Society of Medical Imaging Technology (JAMIT) [11]. The contest was held for the purpose of facilitating machine learning for liver cancer detection. The JAMIT dataset contains 43 CT images (3D volumes) for metastatic liver cancer. Note that each image (volume) contains one or more metastatic liver cancer lesions. We use these images as well as 20 CT images for metastatic liver cancer, extracted from 3Dircadb [12]. We finally use 106 metastatic liver cancer lesions as the dataset. The image resolution is 512 × 512. The slice thickness is 1.0 mm. The pixel spacing is in the range of 0.61 mm to 0.79 mm.

Lesion Image Generation

The proposed method generates various lesion images by the combination of the following three methods:

Method 1: Synthesis using Poisson Blending [5]
Method 2: Generation based on a CT value distribution [6]
Method 3: Generation using DCGANs [8]

Each method is described below in details.

Method 1: Synthesis Using Poisson Blending

Method 1 synthesizes the image region of metastatic liver cancer with a normal CT image using Poisson Blending [4]. First, the method uses GrabCut [13] in order to interactively extract the lesion area. Next, the method resizes the lesion area for the scale variation of the lesion, and then overlays the lesion area (foreground region) on a normal CT image (background region) at a possible position. Here, the density values in the foreground region may be largely different from those in the background one, which leads to the discontinuity of the density value distribution in the synthesized image. Considering that point,

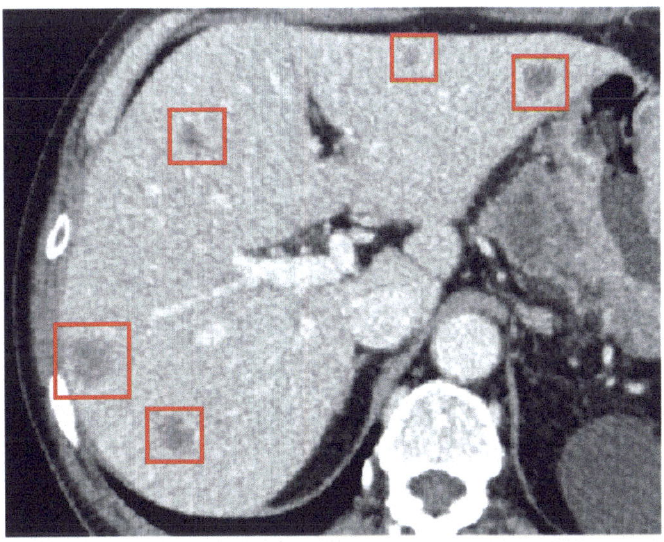

Fig. 2 Example of metastatic liver cancer in a CT image (the rectangles indicate lesion regions)

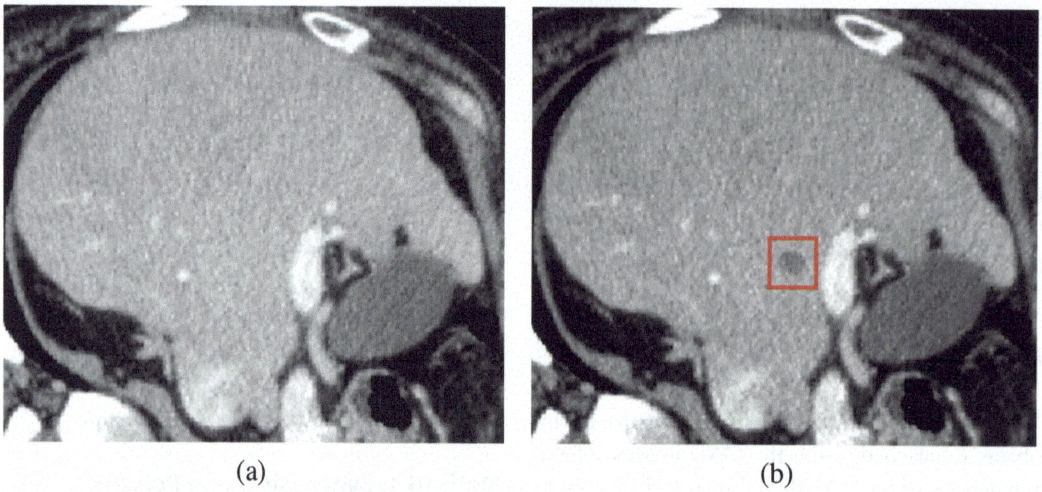

Fig. 3 Example of a lesion image generated by Method 1. (**a**) Original image. (**b**) Synthetic image

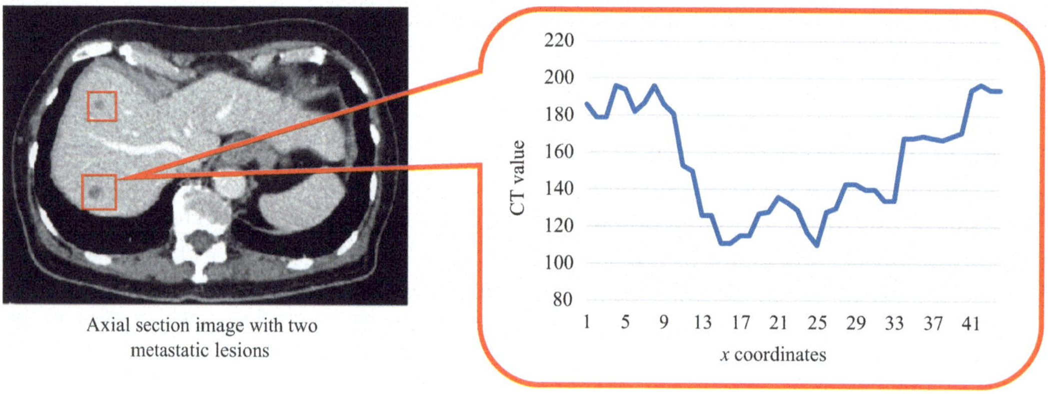

Fig. 4 Example of the CT value profile of a metastatic liver cancer region

the method adjusts the density values of the background region in order that the average CT value in the foreground is equal to that in the background. Figure 3 shows an example of an image generated by the method.

Method 2: Generation Based on a CT Value Distribution

The one-dimensional CT value profile of metastatic liver cancer generally has a bathtub-like shape because of its spherical shape, as shown in Fig. 4. Lesion images with such a shape can be simulated by the following function:

$$g(\boldsymbol{p}) = f(\boldsymbol{p}) - \frac{k}{1 + \exp(r - n(r))h(\boldsymbol{p})} \quad (1)$$

$$n(r) = r_0 + N(0, \sigma_1^2) \quad (2)$$

$$h(\boldsymbol{p}) = \lambda + N(0, \sigma_2^2), \quad (3)$$

where $\boldsymbol{p} = (x, y, z)$ is the center coordinates of a shadow (foreground) region, and $g(\boldsymbol{p})$ is the output CT value at $\boldsymbol{p}$. $f(\boldsymbol{p})$ is the original CT value at $\boldsymbol{p}$, and k is a parameter for controlling the contrast between the foreground and the background, and r is the distance between $\boldsymbol{p}$ and the center of the synthesized lesion. $N(0, \sigma^2)$ is a random value following a normal distribution with zero mean and a standard deviation σ. Figure 5 shows an example of a CT value profile simulated by Eq. (1) with $n(r) = 0$.

Method 2 generates lesion images based on Eq. (1). Figure 6 shows an example of a real lesion image and a generated one synthesized with a healthy person's CT image together with their CT value profiles.

Method 3: Generation Using DCGANs

Method 3 uses DCGANs to generate 32 × 32 lesion images from an input 100-dimensional value whose element is in the range of [−1, 1]. The network structure of DCGANs used in Method 3 is shown in Fig. 7, and the loss function of the DCGANs is as follows:

$$\min_{G}\max_{D} V(D,G) = E_{x\sim p_{\mathrm{data}}(x)}\left[\log D(x)\right] + E_{x\sim p_z(z)}\left[\log\left(1 - D(G(z))\right)\right], \tag{4}$$

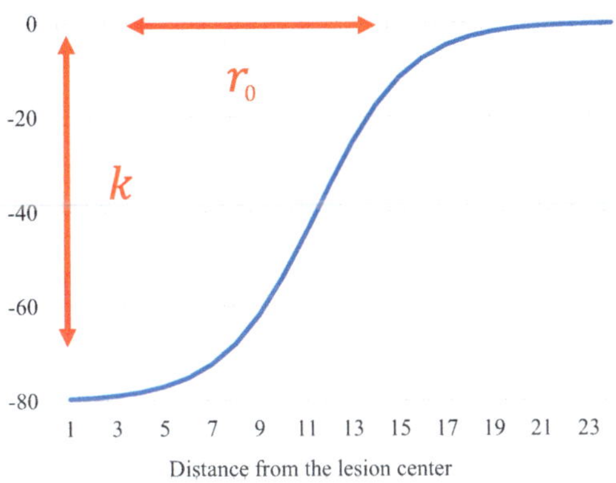

Fig. 5 Example of the CT value profile simulated by Eq. (1) with $n(r) = 0$

where G is a generator, D is a discriminator, and x and z are training data and noise values, respectively. The generator is trained for minimizing the appearance difference between real and synthetic images, and consequently, is expected to generate synthetic images similar to real ones. Figure 8 shows an example of lesion images generated by Method 3.

Selection of the Region of Interest for Lesion Synthesis

Methods 1 and 2 overlay a lesion-like shadow on the region of interest in a real CT image, which enables to generate lesion images similar to real ones. However, in the case of generating a lesion-like shadow near blood vessels, the generated shadow (pseudo-lesion) may be similar to a hemangioma rather than a metastatic liver cancer. Thus, we should first detect blood vessels in a CT image, and then select the region of interest from those sufficiently away from thick blood vessels. We use the multiscale filter [14] in order to extract linear structures based on the Hessian matrices calculated from a liver region. Let I be an image obtained by convoluting a 3D Gaussian filter of a scale s, and the Hessian matrix H at $\boldsymbol{p}$ is described as follows:

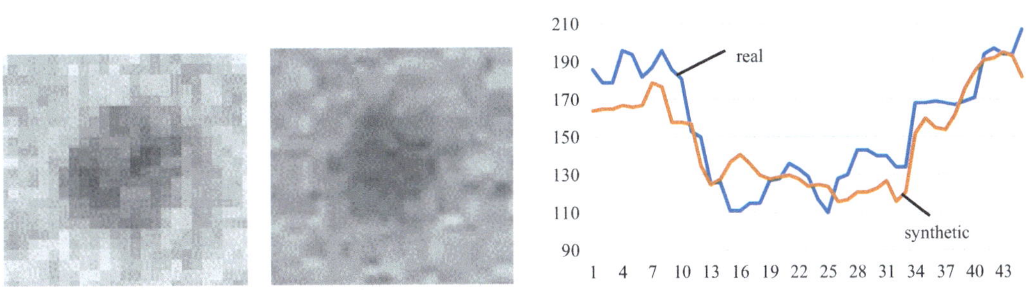

(a) (b) (c)

Fig. 6 Examples of a real lesion CT image and a synthesized one generated by Method 2 ($s_1 = 0$, $H = \lambda$, $k = 80$, and $r_0 = 15$). (**a**) Real image. (**b**) Synthetic image. (**c**) CT value profile

$$H = \begin{pmatrix} I(\boldsymbol{p})_{xx} & I(\boldsymbol{p})_{xy} & I(\boldsymbol{p})_{xz} \\ I(\boldsymbol{p})_{yx} & I(\boldsymbol{p})_{yy} & I(\boldsymbol{p})_{yz} \\ I(\boldsymbol{p})_{zx} & I(\boldsymbol{p})_{zy} & I(\boldsymbol{p})_{zz} \end{pmatrix}, \quad (5)$$

where each element of the matrix is second-order partial derivatives at $\boldsymbol{p}$. Then, the output of the multiscale filter $V_s(\boldsymbol{p})$ is defined by

$$V_s(\boldsymbol{p}) = \begin{cases} 0 & \text{if } \lambda_2 > 0 \text{ } or \text{ } \lambda_3 > 0 \\ \left(1 - \exp(-\frac{R_A^2}{2\alpha^2})\right) \exp\left(-\frac{R_B^2}{2\beta^2}\right) \left(1 - \exp\left(-\frac{S^2}{2\gamma^2}\right)\right) & \text{otherwise} \end{cases},$$

where

$$R_A = \frac{|\lambda_2|}{|\lambda_3|},$$

$$R_B = \frac{|\lambda_1|}{\sqrt{|\lambda_2 \lambda_3|}},$$

$$S = \sqrt{\sum_{j<D} \lambda_j^2}.$$

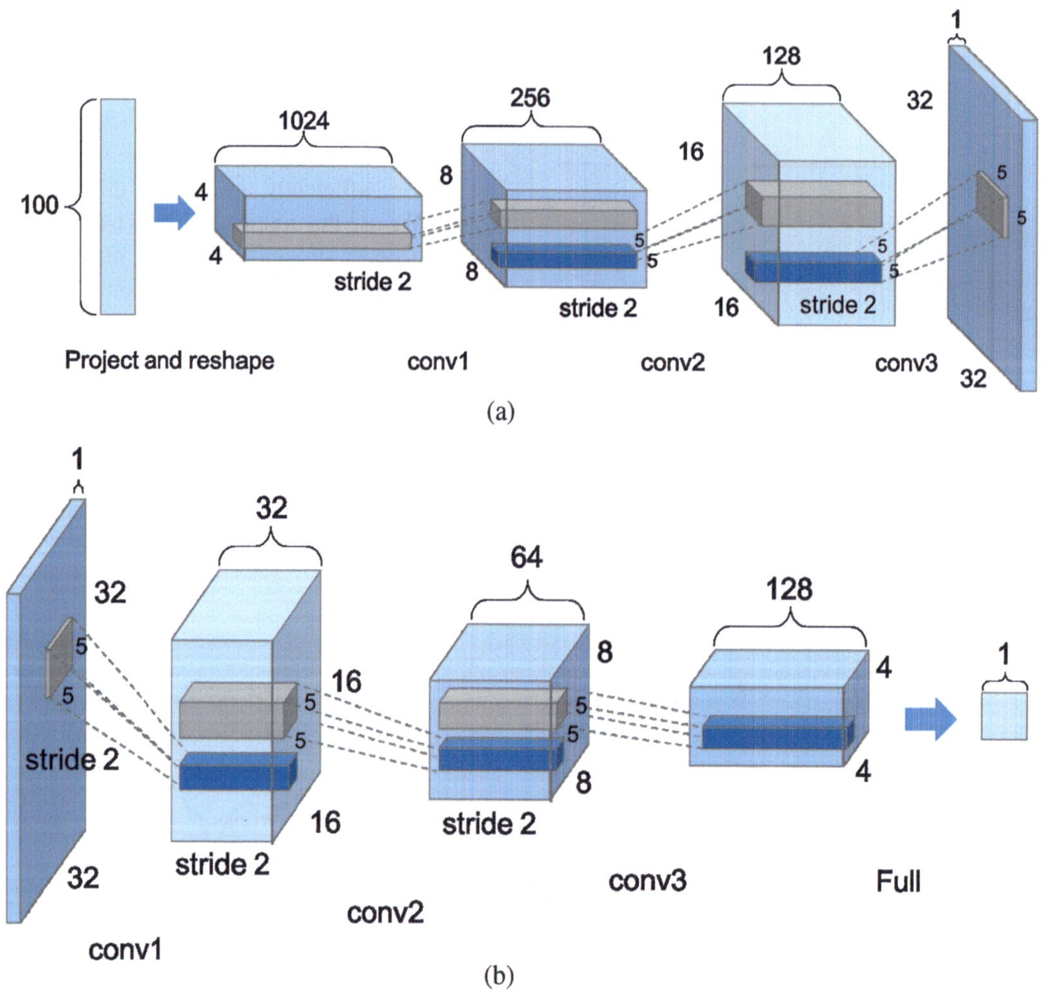

Fig. 7 Network structure of the DCGANs used in Method 3. (**a**) Generator. (**b**) Discriminator

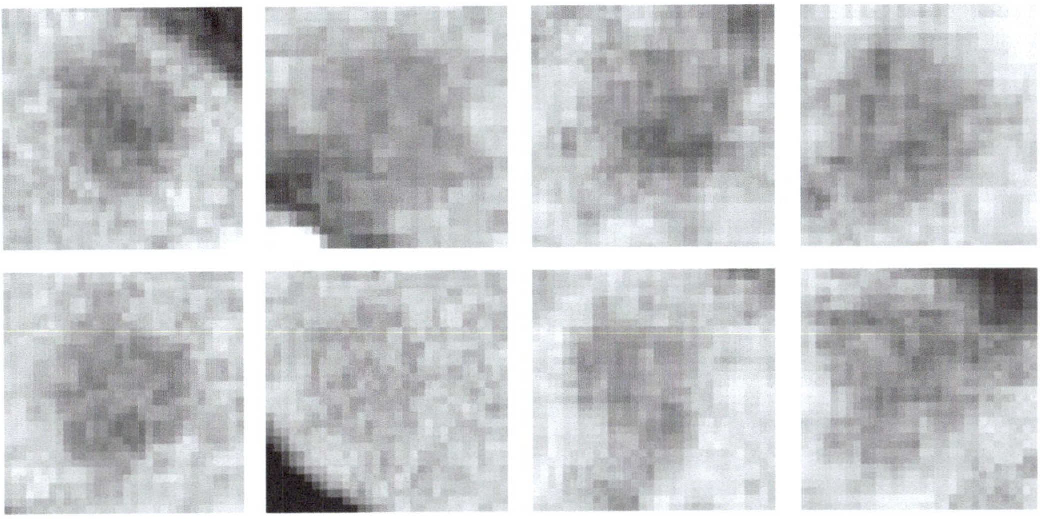

Fig. 8 Examples of synthetic CT images generated by Method 3 using DCGANs in comparison with real ones. (**a**) Real. (**b**) Synthetic

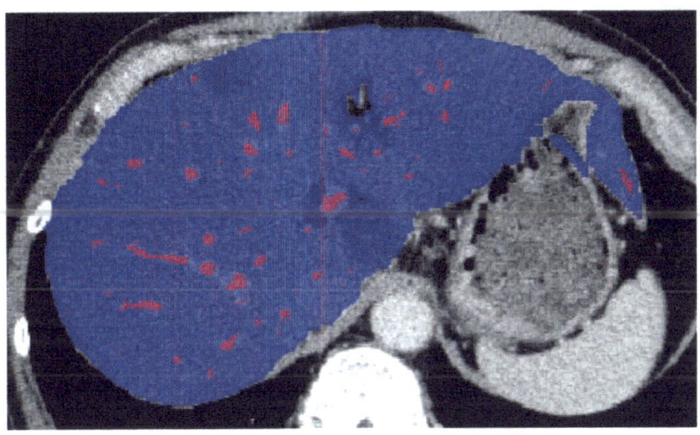

Fig. 9 Example of the detection result of blood vessel regions

Here, λ_i ($i = 1, 2, 3$) is the eigenvalue of the matrix ($|\lambda_1| < |\lambda_2| < |\lambda_3|$). Note that the above-mentioned filter can emphasize CT image pixels whose λ_1 is close to zero and whose λ_2 and λ_3 are large values. In the experiment described later, $\alpha = 5.0$, $\beta = 5.0$, and γ was set to half the value of the maximum Hessian norm. Figure 9 shows an example of a blood vessel region extracted in a liver region. Based on the result for the blood vessel region detection, we can prevent the region of interest from overlapping blood vessels.

Detection Method

Figure 10 shows the detection process flow of the proposed method. In the training step, a 2D CNN and a 3D CNN are both trained using synthetic lesion images generated by the three methods described above in addition to real ones. The detection step is divided into three stages. The first one is candidate region detection by the 2D CNN with a sliding window manner. The second one is the integration of the candidate regions into one volume by using mean shift clustering [15],

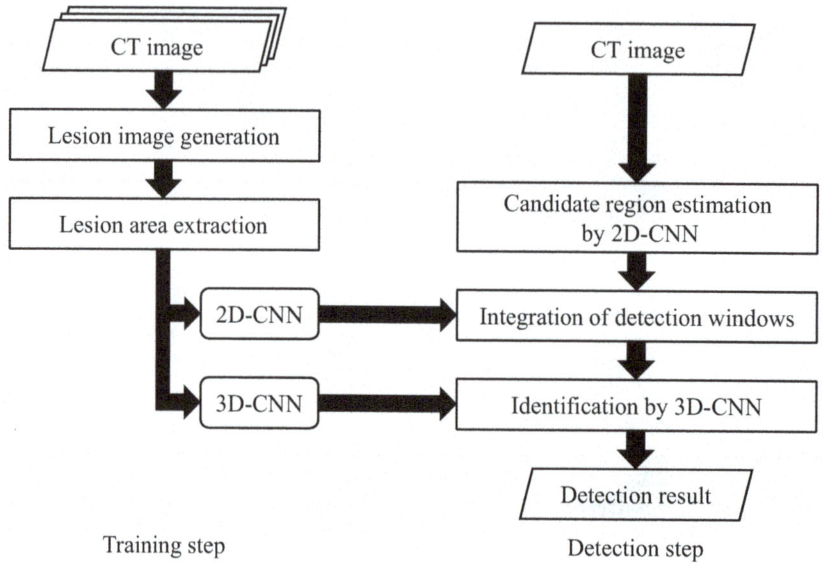

Fig. 10 Detection process flow of the proposed method

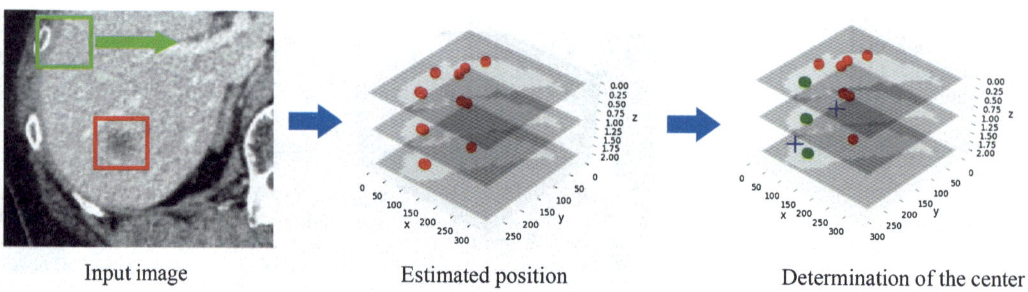

Fig. 11 Integration of detected regions into one volume (red dots indicate detected regions by the 2D CNN, and green dots indicate the integrated volume)

as shown in Fig. 11. The last one is the validation of the resultant regions by the 3D CNN. Here, the network structures of the CNNs are both 8-layer based on VGG [1] as shown in Fig. 12.

Experiments

We evaluated the effectiveness of the proposed method through experiments in terms of the detection rate and the number of false detections per case (FDPC). The training data are shown in Table 1. We used 106 cases of metastatic liver cancer (section "Dataset") for training, and used 20 cases of metastatic liver cancer, which were the evaluation data in the JAMIT dataset [11], for test. Also, We used 32 × 32 and 32 × 32 × 20 lesion CT images for training the 2D CNN and the 3D CNN, respectively. The negative samples for training were randomly extracted from liver regions in non-diseased regions in CT images. For comparative evaluation of the detection accuracy, we constructed six detection models (Models A to F) using synthetic lesion images in addition to real ones, while changing the combination of the three generation methods (section "Lesion Image Generation"). Model A was a comparative method which did not generate lesion images. Models B to E were the modified versions of the proposed method which used one or two of Methods 1 to 3. Model F was the proposed method which used all of Methods 1 to 3. Note that we did not use 3D

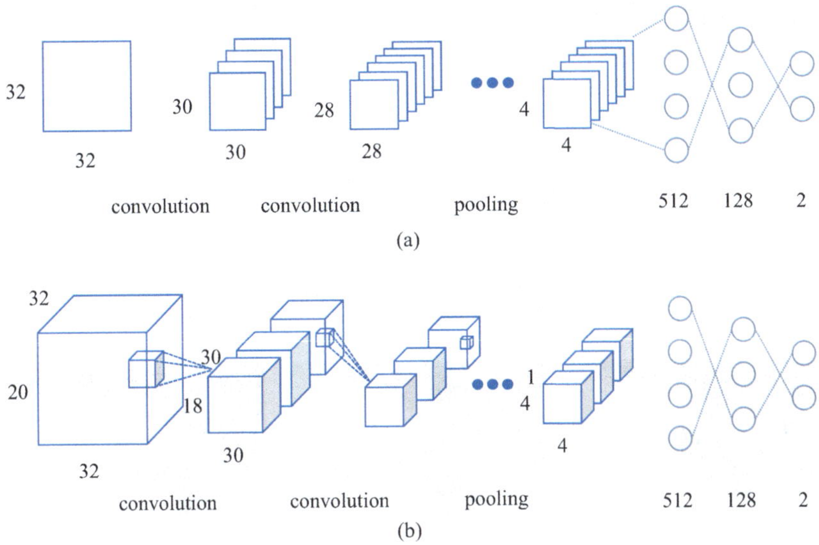

Fig. 12 Network structures of the 2D CNN and the 3D CNN used in the proposed method. (**a**) 2D CNN. (**b**) 3D CNN

Table 1 Training data in the experiments

	Positive samples				
		Synthetic lesion			
Detector	Real lesion	Method 1	Method 2	Method 3	Negative samples
2D CNN	737	3000	3000	3000	4000
3D CNN	106	1000	1000	–	1500

images generated by Method 3 for training the 3D CNN because of too few real images to generate valid synthetic ones.

Results

Experimental results are shown in Table 2. Comparing Model F (the proposed method) with Model A (the comparative one), the proposed method gave 0.30 improvement from 0.65 to 0.95 in terms of the detection rate, and 0.70 improvement from 0.90 to 0.20 in terms of FDPC. The increase of the positive samples would contribute to improve the detection accuracy, because the difference between Models A and F was the number of positive samples for training. The proposed method outperformed all of the other methods. These results showed the effectiveness of the proposed method.

Discussion

We discuss why Model F (the proposed method) achieved the highest detection accuracy among the detection models. Figure 13 shows the detection results obtained by Models A, E, and F. Both Models A and E could not detect the cancer lesion, and moreover, Model E detected one false positive. Model F, which generated lesion images by the combination of the three generation methods, could correctly detect without false positives. With this regard, we confirmed that Method 3 could simulate cancer lesion at the peripheral region of the liver, which can be also seen from Fig. 8. Method 3 generates lesion images by using the DCGANs, which should contribute the improvement of the detection accuracy.

In addition, we analyzed the effectiveness of using the DCGANs from another point of view. We calculated the zero mean normalized

Table 2 Experimental results: detection accuracy

Detection model	Image generation Method 1	Method 2	Method 3	Detection rate	False detections per case (FDPC)
A				0.65	0.90
B	✓			0.75	0.60
C		✓		0.85	0.70
D			✓	0.75	0.50
E	✓	✓		0.90	0.35
F	✓	✓	✓	**0.95**	**0.20**

Bold values show that the values are the detection accuracy obtained by the proposed method (Model F)

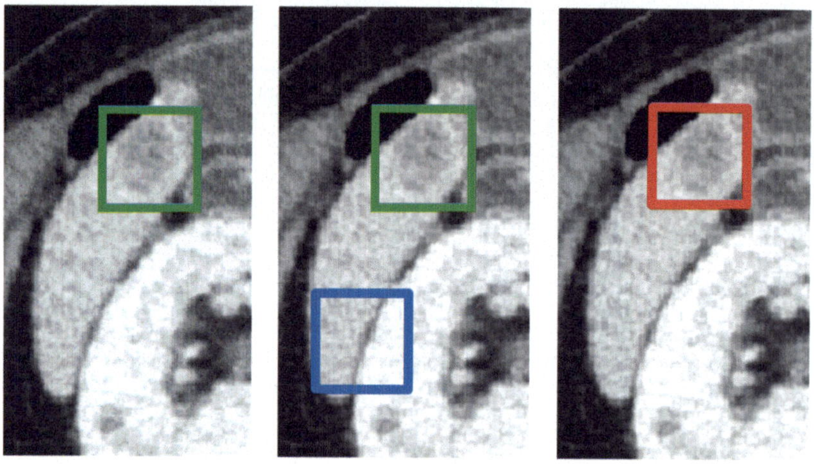

(a) (b) (c)

Fig. 13 Detection results obtained by the detection models: Models A, E, and F (the red, blue, and green rectangles indicate a true positive, a false positive, and a false negative, respectively). (**a**) Model A. (**b**) Model E. (**c**) Model F

cross-correlation (ZNCC) between real lesion images and synthetic ones for each generation method. Figure 14 shows an example of real lesions which could be detected by Model F. The highest ZNCCs between the lesion image and those generated by Methods 1, 2, and 3 were 0.69, 0.62, and 0.86, respectively. This means that a synthetic image generated by the DCGANs was most similar to real one among those generated by Methods 1 and 2, which also indicates the effectiveness of using Method 3.

Examples of the detection results obtained by Model F are shown in Fig. 15. Model F correctly detected the liver cancer region in the image shown in Fig. 15a, and falsely detected at the organ boundary in the image shown in Fig. 15b. This would be because Model F could not sufficiently learn the shape variation of liver cancer. Also, Model F could not detect the liver cancer region in the image shown in Fig. 15c because of the low contrast between the liver and the lesion, which can be improved by introducing contrast enhancement as a preprocessing.

Conclusion

This chapter presented a method to detect metastatic liver cancer from X-ray CT images by a CNN. In order to construct an accurate detector with a small number of training images, we proposed the combination of three kinds of image generation methods: (1) synthesis using Poisson Blending [5], (2) generation based on a CT value distribution [6], and (3) generation using DCGANs [8]. The proposed method detects metastatic liver cancer within a two-stage detection framework: the first one is 2D CNN-based detection as candidate detection, and the second one is 3D CNN-

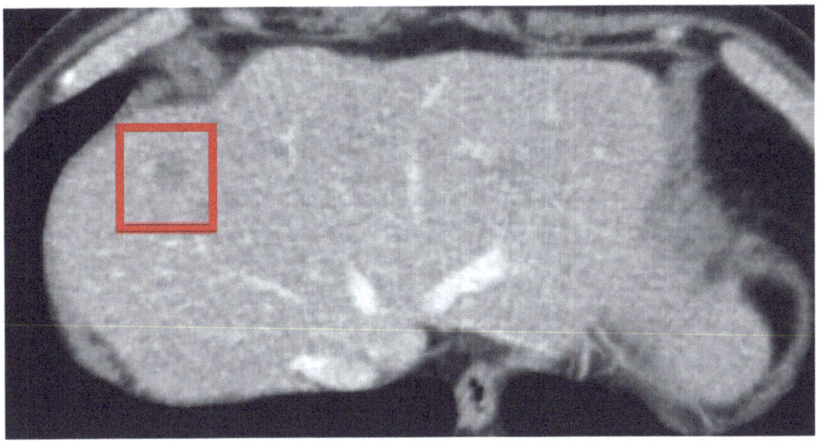

Fig. 14 Example of real lesions which could be detected by Model F

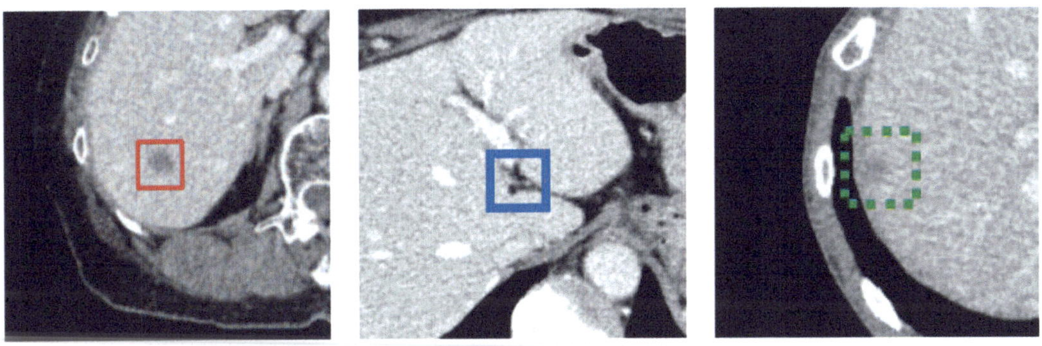

(a) (b) (c)

Fig. 15 Examples of the detection results obtained by Model F. (**a**) True positive. (**b**) False positive. (**c**) False negative

based detection as validation for the detected candidates. Experimental results showed that the proposed method outperformed comparative ones that used one or partial combination of the three generation methods, which indicated the effectiveness of the proposed method.

The future work includes the improvement of the detection accuracy, for example, by introducing contrast enhancement as a preprocessing. Also, we will study another detection frameworks such as region proposal networks.

References

1. Simonyan K, Zisserman A (2015) Very deep convolutional networks for large-scale image recognition. In: Proceedings of the international conference on learning representations
2. Szegedy C, Vanhoucke V, Ioffe S, Shlens J (2016) Rethinking the inception architecture for computer vision. In: Proceedings of the international conference on learning representations
3. Targ S, Almeida D, Lyman K (2016) Resnet in resnet: generalizing residual architectures. In: Proceedings of the international conference on learning representations
4. Prez P, Gangnet M, Blake A (2003) Poisson image editing. In: ACM special interest group on computer, vol 22, no 3, pp 313–318
5. Konishi T, Doman K, Nawano S, Mekada Y (2017) Metastatic liver cancer detection by 3D-CNN using artificial lesion images. In: IEICE technical committee on medical imaging technical research report, vol 116, no 393, pp 21–22 (in Japanese)
6. Konishi T, Doman K, Nawano S, Mekada Y (2017) A study on lesion image synthesis for metastatic liver cancer detection using artificial training images. In: The Japanese society of medical imaging technology, OP3-1 (in Japanese)

7. Goodfellow IJ, Pouget-Abadie J, Mirza M, Xu B, Warde-Farley D, Ozair S, Courville A, Bengio Y (2014) Generative adversarial nets. In: Neural information processing systems, vol 27, pp 2672–2680
8. Mekada Y, Doman K (2018) Learning sample generation for detecting liver cancer using 3D-CNN. In: Proceedings of 40th international engineering in medicine and biology conference, no 370, CThBT19.7
9. Radford A, Metz L, Chintala S (2016) Unsupervised representation learning with deep convolutional generative adversarial networks. In: Proceedings of the international conference on learning representations
10. Christ PF et al (2016) Automatic liver and lesion segmentation in CT using cascaded fully convolutionalneural networks and 3D conditional random fields. In: International conference on medical image computing and computer-assisted intervention. LNCS, vol 9901. Springer, Heidelberg, pp 415–423
11. About CAD contest, http://www.jamit.jp/meetinginfo/cad.html (in Japanese)
12. 3Dircadb, https://www.ircad.fr/research/3dircadb/
13. Wu W, Wu S, Zhou Z, Zhang R, Zhang Y (2017) 3D liver tumor segmentation in CT images using improved fuzzy C-means and graph cuts. In: BioMed research international, vol 2017, pp 1–11
14. Frangi AF, Niessen WJ, Vincken KL, Viergever MA (1998) Multiscale vessel enhancement filtering. In: Proceedings of medical image computing and computer assisted intervention society, pp 130–137
15. Cheng Y (1995) Mean shift, mode seeking, and clustering. IEEE Trans Pattern Anal Mach Intell 17(8):790–799

Retinopathy Analysis Based on Deep Convolution Neural Network

Yuji Hatanaka

Abstract

At medical checkups or mass screenings, the fundus examination is effective for early detection of systemic hypertension, arteriosclerosis, diabetic retinopathy, etc. In most cases, ophthalmologists and physicians grade retinal images by the condition of the blood vessels, lesions. However, human observation does not provide quantitative results, thus blood vessel analysis is an important process in determining hypertension and arteriosclerosis, quantitatively. This chapter describes the latest automated blood vessel extraction using the deep convolution neural network (DCNN). Diabetic retinopathy is a common cardiovascular disease and a major factor in blindness. Therefore, early detection of diabetic retinopathy is very important to preventing blindness. A microaneurysm is an initial sign of diabetic retinopathy, and much research has been conducted for microaneurysm detection. This chapter also describes diabetic retinopathy detection and automated microaneurysm detection using the DCNN.

Keywords

Hypertensive retinopathy · Diabetic retinopathy · Cardiovascular disease · Retinal image · Fundus examination

Introduction

A fundus examination is carried out at mass screenings and medical checkups, and is effective for early detection of systemic hypertension, arteriosclerosis, diabetic retinopathy, glaucoma, etc. The observation points of a fundus examination are the condition of the blood vessels, retina, and macula, and the shape of the optic disc. For many years, ophthalmologists and physicians classified retinal images as indicating hypertension and arteriosclerosis using the Scheie classification [1], which is shown in Table 1. Meanwhile, for a number of patients, serious signs can be reduced by controlling hypertension. Because mild signs do not have enough basis, physicians and ophthalmologists cannot give concrete advice, such as the significance of controlling hypertension, to patients. Wong and Mitchell proposed a novel classification based on a cohort study that established a connection with grades based on retinal signs and risk factors for cardiovascular disease [2]. The Wong–Mitchell classification is shown in Table 2.

Y. Hatanaka (✉)
University of Shiga Prefecture, Hikone-city, Japan
e-mail: hatanaka.y@e.usp.ac.jp

G. Lee, H. Fujita (eds.), *Deep Learning in Medical Image Analysis*,
Advances in Experimental Medicine and Biology 1213,
https://doi.org/10.1007/978-3-030-33128-3_7

Table 1 Scheie classification

	Hypertension (H)	Arteriolosclerosis (S)
Grade I	Diffuse arteriolar narrowing	Broadening of arteriolar light reflex
Grade II	Grade I + Focal arteriolar constriction	Grade I + Arteriovenous nicking
Grade III	Grade II + Hemorrhages, Exudates	Copper wire arteries
Grade IV	Grade III + Papilledema	Silver wire arteries

Table 2 Wong–Mitchell classification

	Retinal sign	Systemic association
Mild	Focal/general arteriolar narrowing, Copper wiring, Arteriovenous nicking	Mild association with risk of clinical stroke and cardiovascular mortality
Moderate	Hemorrhages, Hard exudate, Cotton wool spots, Microaneurysm	Strong association with clinical stroke, cardiovascular mortality and cognitive decline
Malignan	Other signs + optic disc edema	Strong association with death

Ophthalmologists and physicians often grade retinal images by subjective judgment based on visual examination. Therefore, the automated detection of abnormalities and quantitative measurement of tissue have been proposed. In recent years, deep learning techniques were applied to retinal analysis methods. This chapter describes retinal analysis for (1) general arteriolar narrowing detection in hypertension and (2) microaneurysm detection in diabetic retinopathy using deep learning techniques.

General Arteriolar Narrowing Detection

Focal or general arteriolar narrowing is on retinal image is a marker of microvascular damage from hypertension. General arteriolar narrowing is determined by the arteriovenous ratio (AVR), which has shown to be a useful marker in vascular disease. The smaller the AVR is, the more serious the general arteriolar narrowing becomes, because the narrower the focal caliber of an artery, the more serious the condition of the artery. AVR classifies a case to normal (over 2/3), grade 1 (2/3 to 1/2), grade 2 (1/2 to 1/3), and grade 3 (under 1/3). However, this grading is only a guide. Note that a normal AVR is 2/3 or 3/4, because the differences between graders would be most likely due to physician clinical experience and University training. To detect general arteriolar narrowing, automated measurement methods of the artery caliber have been developed. This consists of three processes: (1) blood vessel extraction, (2) automated classification of the artery and vein, and (3) measurement of the blood vessel caliber.

Blood Vessel Extraction

Related Works

Many methods have been established for blood vessel extraction. Most of methods were used vessel enhanced filter or machine learning. Staal et al. proposed a method based on image ridge extraction using a k-nearest neighbor (kNN) classifier with properties of the patches and line elements [3]. Soares et al. proposed a method based on a two-dimensional Gabor wavelet transform [4]. Rangayyan et al. proposed a method using the design of a bank of directionally sensitive Gabor filters [5]. Ricci et al. proposed a method using two orthogonal line detectors along with the gray level of the target pixel [6]. Akasaka et al. proposed a method based on ensemble learning [7]. Various authors have also proposed simple methods using a double-ring filter [8], black top-hat transformation [9], and a combination of these [10].

In general, determining parameters, such as optimal filter size and shape, is difficult. A general filtering technique depends strongly on a preset model. The patterns of blood vessels are so varied that it is not practical to design a flexible blood vessel model. Therefore, a method has been proposed based on high-order local autocorrelation (HLAC) features, which are shift-invariant and model-free, and are appropriate for center-shifted hotspot pattern feature extraction [11].

After approximately 2015, deep learning techniques were applied to blood vessel extraction. Fu et al. proposed a method based on combining four parallel convolutional layers [12]. Martina et al. proposed a method using a deep convolutional neural network (DCNN) structured with four convolutional layers and four pooling layers [13]. Dasgupta et al. proposed a method using a fully convolutional network (FCN) [14]. A method was also proposed using a simple patch-based DCNN [15]. The method consisted of the preprocessing and classification of blood vessels and non-vessels. This chapter describes this simple method.

Database

The retinal images used in this study were obtained from the Digital Retinal Images for Vessel Extraction (DRIVE) database [3]. This database includes 40 retinal fundus images that were obtained from a diabetic retinopathy screening program in the Netherlands and are equally divided into training and testing sets. The images were 565 × 584 pixels in size, with 24-bit color. For each image, the manual segmentation result for blood vessels is provided as a reference standard.

Preprocessing

Non-uniformly illuminated images, such as retinal images, exhibit local luminosity and contrast variability. This problem could affect a DCNN, thus global contrast normalization (GCN) [16] is applied in general. GCN is based on an estimation of the luminosity and contrast variability in the background part of the image, and the subsequent compensation of this variability in the whole image.

The pixels with a high correlation of the adjacent pixels should be removed for the DCNN. Zero component analysis (ZCA) whitening [17], principal component analysis (PCA) whitening [18], etc. are well-known preprocessing techniques for DCNNs. A whitening is a linear algebra operation that reduces the redundancy in neighboring pixels. Image features are enhanced by the reduction in redundancy. PCA is global in space and local in frequency, and ZCA is local in space and global in frequency [19]. Therefore, PCA whitening works like a high pass filter. In the patch image, the blood vessel region is low frequency, thus ZCA is better. However, whitening tends to enhance the choroid layer behind the retina. Therefore, the patch using the green channel of the color retinal image is input to the DCNN.

In order to choose the preprocessing method, LeNet [20] was tested as the DCNN architecture. LeNet is a well-known DCNN architecture. In this test, LeNet was trained using each of the 400,000 blood vessel patches and non-vessel patches of 27 × 27 pixels in the test images of the DRIVE database. The preprocessing candidates were (1) GCN + ZCA, (2) the green channel image, and (3) the green channel image + median filter, as shown in Fig. 1. The performances were compared using the area un-

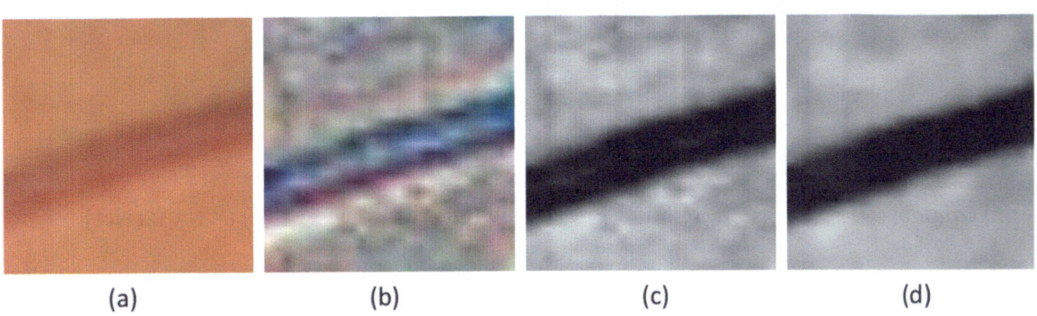

Fig. 1 Examples of preprocessed images. (**a**) Original. (**b**) GCN + ZCA. (**c**) Green channel. (**d**) Green + median

der the curve (AUC) of the receiver operating characteristic (ROC) analysis. The AUCs of (1)–(3) were 0.936, 0.948, 0.939, respectively. The GCN + ZCA image shows high contrast. However, the GCN + ZCA enhanced reflection region, so that the AUC of this method was less than that of green channel image. The median filter removed irregular color, as shown in Fig. 1d, but it did not contribute to the classification of vessels and non-vessels. Therefore, the green channel image represented better preprocessing for blood vessel extraction using a DCNN, contrary to the author's expectations.

Blood Vessel Extraction Using DCNN

Figure 2 shows the workflow of blood vessel extraction. The green channel of the color retinal image was extracted, and patches of $k \times k$ pixels were extracted for the DCNN. In this section, two simple architectures, LeNet and AlexNet [21], were compared. LeNet consists of a convolutional layer followed by a pooling layer, another convolution layer followed by a pooling layer, and then two fully connected layers similar to the conventional multilayer perceptrons. We use a slightly different version from the original LeNet implementation, replacing the sigmoid activations with rectified linear unit (ReLU) activations for the neurons. Moreover, AlexNet contained eight layers; the first five were convolutional layers, some of them followed by max-pooling layers, and the last three were fully connected layers. For the training of LeNet and AlexNet, 800,000 patches of 21 × 21 pixels were randomly selected from the green channel of the color images. The AUCs of LeNet and AlexNet were 0.924 and 0.889, respectively. The pooling layer presented the problem of location invariance so that the blood vessel region tended to get over-extracted when the number of max-pooling layers was increased.

The patch size is an important parameter of the DCNN. As a preliminary examination, patches of 11 × 11, 21 × 21, and 27 × 27 pixels were tested. The AUCs were 0.943, 0.924, and 0.921, respectively. Therefore, the patch of 11 × 11 pixels was the best.

Finally, the outputs of the DCNN were binarized using a threshold value with the highest accuracy, and the small regions were assumed to be noise, and removed. An example of blood vessel extraction is shown in Fig. 3. The blood vessels around the optic disc were extracted well, as shown in Fig. 3d. Although no peripheral blood vessels were extracted, they are unnecessary to diagnosing hypertension.

Table 3 shows the performance of 11 blood vessel extraction methods. The LeNet-based method [15] is better than the previous feature-based methods. The FCN-based method [14] is the best of these methods. LeNet is a type of classification neural network, so segmentation is needed for the binarization process. However, FCN uses semantic segmentation, and so a binarization process is unnecessary. If there are enough training data, FCN will perform better than a classification neural network.

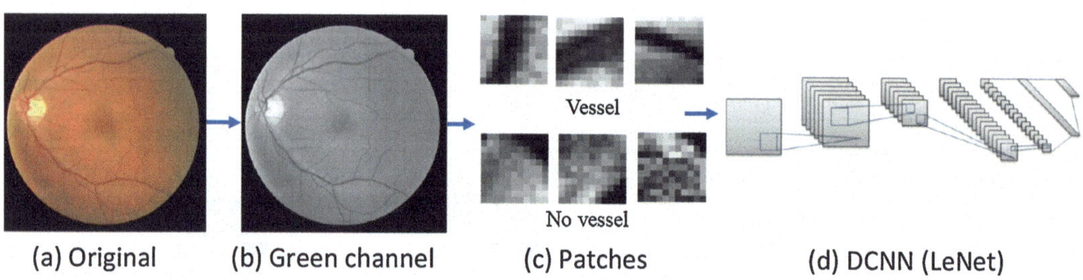

Fig. 2 The flow of blood vessels extraction. (**a**) The retinal images were 565 × 584 pixels in size, with 24-bit color. (**b**) The color image is changed to gray scale one using the green channel pixel values. (**c**) All patches in the gray-scale image are classified into blood vessels and non-vessels by using DCNN (**d**). The patch sizes are 11 × 11 pixels in DRIVE and 17 × 17 pixels in INSPIRE-AVR

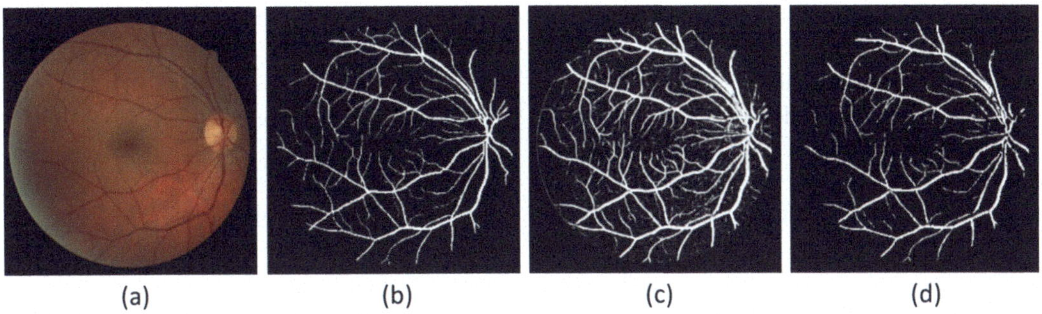

Fig. 3 Examples of blood vessels extraction. The output of CNN (**c**) looks like binarized one, because almost output values of DCNN were under 0.10 or over 0.9. (**a**) Original. (**b**) Ground truth. (**c**) Output of DCNN. (**d**) Binarized image

Table 3 Comparison of blood vessel extraction

Method	Sensitivity	Specificity	Accuracy	AUC
Feature-based methods:				
Human observer	0.776	0.973	0.947	N/A
Staal [3]	0.719	0.977	0.944	0.952
Soares [4]	N/A	N/A	0.947	0.961
Rangayyan [5]	N/A	N/A	N/A	0.961
Ricci [6]	N/A	N/A	0.960	0.963
Akatsuka [7]	0.877	0.942	N/A	N/A
Hatanaka [11]	N/A	N/A	0.945	0.960
Deep learning-based methods:				
Fu [12]	0.715	0.977	0.943	N/A
Martina [13]	0.707	0.980	0.945	0.959
Dasgupta [14] (FCN)	0.769	0.980	0.953	0.974
Ikawa [15] (LeNet)	0.700	0.980	0.944	0.964

Detection of Arteriolar Narrowing Using AVR

Related Works

Niemeijer et al. proposed an AVR measurement method by classifying arteries and veins based on color pixels in the centerlines of the blood vessels [22]. Dashbozorg et al. proposed a method correcting the graph representing the vessel tree by applying rules based on a priori knowledge of the retinal blood vessels [23]. Vázquez et al. [24] proposed a classification method based on a minimal path approach, where the vessels are segmented, measured, and classified according to several circumferences concentric to the optic disc. Authors have also reported the grading of hypertensive changes based on measurements of the diameter of arteries and veins [25]. In this system, the arteries and veins are classified according to linear discriminant analysis (LDA) of eight pixel-based features at the centerline of the candidate blood vessel. However, this approach was not sufficient for cases with close contact between the artery of interest and an imposing vein. Therefore, additional steps for accurate segmentation of arteries and veins, which were not identified using the previous method [25], have been added to better identify major veins in the red channel of a color image [26]. To identify major arteries, a decision tree with three features was used.

In order to jointly segment and classify vessels as arteries and veins, an FCN was applied for AVR measurement [27, 28]. Inces et al. presented an approach based on an FCN that was specifically adapted to artery/vein classification [27]. The performance of the method was evaluated on the RITE dataset [29], and the accuracy was 96% on thick vessels with an overall accuracy of 84%. Girarda et al. also proposed a method based on an FCN [28]. Their method involved two steps: initial semantic segmentation using a CNN model, followed by propagation of the CNN output. The method achieved an accuracy of 94.8% for vessel segmentation. The A/V classification achieved a specificity of 92.9% with a sensitivity of 93.7% on the DRIVE database. A method using a CNN for the simplification of the arteriovenous classification process has also been proposed [15].

On the INSPIRE-AVR database [22], this method classified 90% of blood vessels as arteries and veins. This chapter describes the classification of arteries and veins for AVR measurement.

Database

The INSPIRE-AVR database [22] was used for the evaluation of the arteriovenous classification method. This database contains 40 retinal images at a resolution of 2392 × 2048 pixels, and the AVR values were added manually by two experts. The vessel centerlines and the vessel types labeled by Dashtbozorg et al. [23] were also used.

Classification of Arteries and Veins

The previous methods [27, 28] jointly extracted and classified arteries and veins using an FCN. However, this method extracted and classified them separately.

The image resolution of the images in INSPIRE-AVR is higher than those in DRIVE. Therefore, the images were downsized to one-quarter size (1196 × 1024 pixels), and the blood vessels were extracted using LeNet, which is described in section "Blood Vessel Extraction Using DCNN." The patch size of LeNet was 17 × 17 pixels. The binarized image was finally resized to the original resolution.

The arteries and veins were then classified by a patch-based DCNN. The difference in pixel value between arteries and veins is small, so the original color image was used for the DCNN without normalization.

In order to determine the DCNN architecture, AlexNet and GoogLeNet [30] were preliminarily evaluated using the RITE database [29]. Both AlexNet and GoogLeNet were trained using 25,738 artery patches and 25,872 vein patches. At the center of each patch there were blood vessels labeled as artery or vein. The AUCs of AlexNet and GoogLeNet were 0.601 and 0.620, respectively. Therefore, AlexNet was applied to the classification of arteries and veins. In the preliminary examination, the AUC was increased by data augmentation. For data augmentation gamma correction and random erasing were used on each AUC to increase them to 0.701 and 0.713, respectively. However, the number of patches in INSPIRE-AVR was high enough that data augmentation was unnecessary.

The AVR was measured using the Knudtson et al. method [31]. The AVR is generally determined using the six largest arteries and veins on retinal images centered at the optic disc. The AVR measurement zone is the ring region centered on the optic disc, which is the region between twice the optic disc diameter to three times the optic disc diameter, as shown in Fig. 4. Only the blood vessel branches in the AVR measurement zone were evaluated in this method. The method used

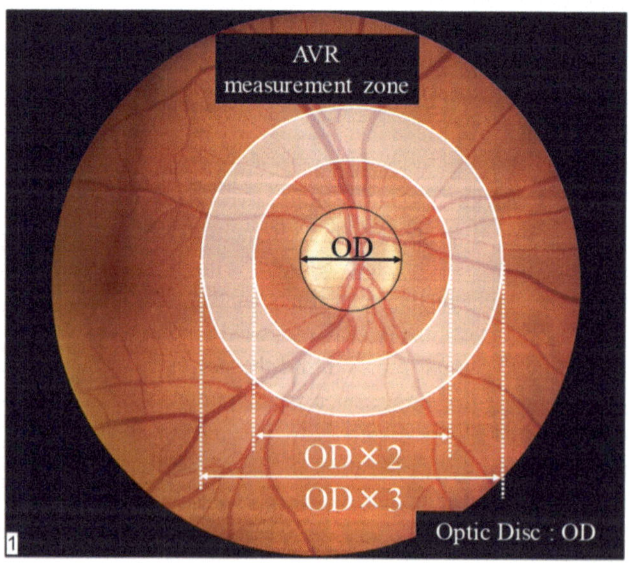

Fig. 4 AVR measurement zone. AVR is calculated using each larger six arteries and veins in this zone, which is the ring region centered on the optic disc, which is the region between twice the optic disc diameter to three times the optic disc diameter

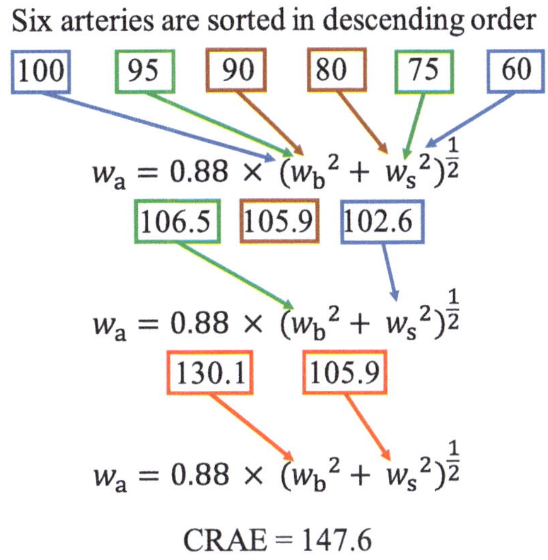

Fig. 5 An example of CRAE calculation. In the top line, blue boxes show the highest value and the lowest one pair. "w_a" is defined in Eq. (2). w_a is 102.6 by calculation using two blue box values

Table 4 Results of classified arteries and veins in AVR measurement zone

	Artery	Vein
Number of vessel branches	84/88	64/81
Success ratio	0.955	0.790
Mean of success ratio	0.878	

Table 5 Results of the six largest classified arteries and veins in the AVR measurement zone

	Artery	Vein
Number of vessel branches	56/60	56/60
Success ratio	0.933	0.933
Mean of success ratio	0.933	

AlexNet to evaluate INSPIRE-AVR images and labeled data [23]. The size of the patches in the RGB color images and green images was 41 × 41 pixels. The green channel contributed strongly to the color images, so that patches could not be classified using DCNN based on the RGB images and were correctly classified using DCNN based on the green images. Note that diameters greater than 15 pixels were used for wider blood vessels. When the method was evaluated using four-fold cross validation, the AUC, accuracy, precision, recall, and f-value reached 0.930, 0.871, 0.867, 0.896, and 0.881, respectively. However, patches classified as artery and vein were mixed in a certain vessel branch. Therefore, a certain vessel branch was classified into artery and vein based on the majority rule using classified patches. The results of the classified arteries and veins are shown in Table 4. The method tends to wrongly classify narrow veins. The success ratios for the six largest arteries and veins are shown in Table 5. 93% of arteries and veins were classified, so the method was proven effective.

AVR Measurement

The AVR is the ratio of the central retinal artery equivalent (CRAE) and central retinal vein equivalent (CRVE),

$$\mathrm{AVR} = \frac{\mathrm{CRAE}}{\mathrm{CRVE}} \tag{1}$$

where CRAE and CRVE are defined in Eqs. (2) and (3).

$$w_a = 0.88 \times \left(w_b{}^2 + w_s{}^2\right)^{1/2} \tag{2}$$

$$w_v = 0.88 \times \left(w_b{}^2 + w_s{}^2\right)^{1/2} \tag{3}$$

where w_b and w_s are the wider artery and the narrower artery, respectively. An example of CRAE calculation is shown in Fig. 5. Six arteries are sorted in descending order and divided into high and low groups. The highest and lowest values in each group are selected, and the arteriole diameter is calculated using Eq. (2). In the same way, the second arteriole diameter is calculated using

Table 6 Results of AVR measurement

	Obs. 1	Obs. 2	Niemeijer [22]	DCNN
Mean	0.67	0.66	0.67	0.66
Standard deviation	0.08	0.08	0.07	0.09
Maximum value	0.93	0.85	0.81	0.89
Minimum value	0.52	0.45	0.55	0.48

Table 7 Error for observer 1

	Obs. 2	Niemeijer [22]	DCNN
Mean	0.05	0.08	0.06
Standard deviation	0.04	0.06	0.05
Maximum value	0.29	0.21	0.23
Minimum value	0.00	0.00	0.00

the second highest and lowest values, and the third arteriole diameter is also calculated. Then, two arteriole diameters are calculated using three arteriole diameters. Finally, the CRAE is calculated using two arteriole diameters. CRVE is also calculated in the same way as CRAE.

In INSPIRE-AVR, two observers manually measured AVR. Table 6 shows the results of AVR measurement by the two observers, using the Niemeijer et al. method [22] and this method. The error for observer 1 was calculated, as shown in Table 7. All results were similar, so the method based on the DCNN was shown to work effectively. The method tended to extract a value wider than the real value of vessels with a caliber of less than 15 pixels, thus narrow vessels should be extracted using the DCNN trained on narrow vessel patches.

INSPIRE-AVR did not classify into the abnormal and normal cases. In this section, it was defined that the case with over 2/3 AVR is normal, and another case is abnormal. If AVRs of observer 1 are ground truth, this database includes 22 abnormal cases and 18 normal ones. The sensitivity, specificity, and accuracy used DCNN were 0.77 (17/22), 0.78 (14/18), and 0.78 (31/40), respectively. On the other hands, the sensitivity, specificity, and accuracy of observer 2 were 0.82 (18/22), 0.56 (10/18), and 0.70 (28/40), respectively. Observer 2 tended to underestimate AVR than observer 1, thus 8 normal cases based on AVRs observer 1 were classified abnormal using AVRs by observer 2. From the above results, DCNN could measure AVR quantitatively and could classify into abnormal and normal like human observer.

Microaneurysm Detection

Diabetic retinopathy is a leading cause of vision loss [32]. Early detection and treatment of diabetic retinopathy is critical to prevent vision loss. It is important to detect microaneurysm early because it is an early symptom of diabetic retinopathy. In screenings and periodical checkups, non-contrast retinal images are used. However, microaneurysms appear as small dark dots in a retinal image. Therefore, microaneurysm detection in non-contrast retinal images is challenging and difficult work.

Related Work

The Retinopathy Online Challenge competition, which was focused on microaneurysm detection, was held at SPIE 2009 [33]. The database consists of 50 training images and 50 test images, and microaneurysms marked by four people. The result of one expert was a sensitivity of 0.49 and 1.08 false positives per image [33]. The precision and f-values were 0.76 and 0.59, thus microaneurysm detection on this database was very difficult. Niemeijer et al. also provided a comparison of five methods using this database [33].

Several research groups have been developing automated microaneurysm detection methods using retinal images [34–39]. Adal et al. proposed a microaneurysm detection method using two eigenvalues based on the Hessian matrix [34], and Antal et al. proposed a method based on an ensemble of microaneurysm detectors [35]. Seoud et al. proposed a method based on dynamic shape features [36], whereas Dai et al. proposed a method based on density gradient vector analysis [37]. A method based on gradient vector concentration has also been proposed [38], as well as on combining three microaneurysm

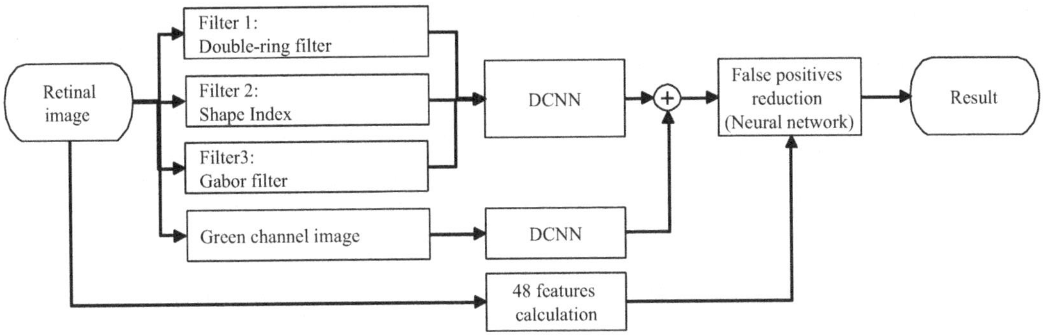

Fig. 6 Detection of microaneurysms using DCNN. The initial microaneurysm candidates are determined by the bitwise and of two DCNNs. Final microaneurysms are determined by false positives reduction based on neural network, which is 3-layer perceptron, with texture and shape features

detectors: a double-ring filter, Gabor filter, and shape index [39]. These previous methods had many parameters to set. However, the DCNN was a breakthrough technique in object classification and pattern recognition. Haloi et al. proposed a method using deep learning with dropout using a maxout function [40]. Chudzik et al. proposed a method based on U-Net using fine tuning [41]. A method using GoogLeNet and feature analysis has also been proposed [42], and this method describes in this section.

Database

The Retinopathy Online Challenge database [33] is quite well known and well used. However, the ground truth in the test images is not available, so researchers use the Standard Diabetic Retinopathy Database Calibration level 1 (DIARETDB1) [43], E-OPHTHA [44]. This section describes DIARETDB1. This database consists of 28 training images with 83 microaneurysms and 61 test images with 100 microaneurysms. The image resolution is 1500 × 1152 pixels. All microaneurysms were determined by four experts.

Methods

This method involved preprocessing, microaneurysm detection using the DCNN, and reduction of false positives using 48 image features, as shown in Fig. 6. In general, a normal image or an image with applied whitening is input to the DCNN. Otherwise, the DCNN learns image features during the training process. However, a microaneurysm is very small and low contrast, so the DCNN cannot learn the microaneurysm feature. The ZCA cannot enhance microaneurysm effectively either. Therefore, the microaneurysm is enhanced using specific filters. The DCNN also detects many false positives because microaneurysm detection is very difficult. Therefore, another DCNN also detected microaneurysms, and the candidate detected by both DCNN was determined to be microaneurysm candidates. Finally, the false positives were removed using 48 image features.

Preprocessing

The contrast between a microaneurysm and the retinal area is highest in the green channel of a color image. Therefore, we obtained green channel component images for blood vessel extraction and microaneurysm detection. Differences exist in the contrast of microaneurysms in retinal images. To reduce the adverse effects of these differences on image processing, a double-ring filter, Gabor filter, and shape index based on the Hessian matrix, which were used in a previous method [39], were applied during preprocessing.

The double-ring filter enhances a microaneurysm candidate region using the difference between the outer-ring region and the inner-circle region, as shown in Fig. 7.

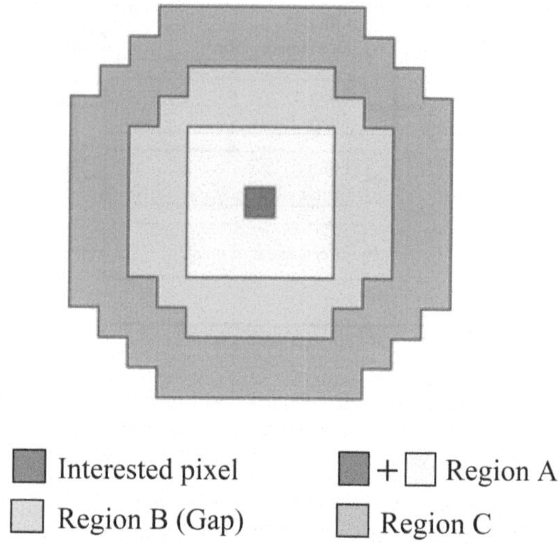

Fig. 7 Structure of double-ring filter. Region A means a microaneurysm region, region C means around of microaneurysm region, and region B means a gap of region A and B

$$O(i, j) = 128 - (\mathrm{TH}(i, j) - A_{\mathrm{mean}}(i, j)) \quad (4)$$

where TH(i, j) is the threshold value determined with the p-tile method in an outer-ring region (region C), and $A_{\mathrm{mean}}(i, j)$ is the mean of the pixel values in an inner-circle region (region A).

The Hessian matrix is a square matrix of functions derived by second-order partial differentiation. Shape index S is defined using two eigenvalues ($\lambda_1 \geq \lambda_2$) of the Hessian matrix.

$$S = \begin{cases} -\frac{2}{\pi}\tan^{-1}\frac{(\lambda_1+\lambda_2)}{\lambda_1-\lambda_2} & (\lambda_1 \neq \lambda_2) \\ -1 & (\lambda_1 = \lambda_2 > 0) \\ 1 & (\lambda_1 = \lambda_2 < 0) \end{cases} \quad (5)$$

where $S = -1$ means that the distribution is cup shaped. The pixel distribution of a typical microaneurysm is similar to a cup.

The Gabor filter enhances a specific frequency. It is represented as follows:

$$g(x, y) = \exp\left(-\frac{x'^2 + \gamma^2 y'^2}{2\sigma^2}\right)\cos\left(2\pi\frac{x'}{\lambda}\right) \quad (6)$$

where

$$x' = x\cos\theta + y\sin\theta,\ y' = -x\sin\theta + y\cos\theta \quad (7)$$

where λ is the wavelength, γ is the aspect ratio, σ is the standard deviation of the Gaussian function, and θ is the angle. The filtered image takes the maximum value of $g(x, y)$ by setting $\theta = 0, \pi/12, 2\pi/12 \ldots, 11\pi/12$. The sizes of the microaneurysms also vary; thus, three values of the wavelength $\lambda = \lambda_a, \lambda_b, \lambda_c$ are defined, and their outputs are $G_a(i, j)$, $G_b(i, j)$, $G_c(i, j)$, respectively. In the microaneurysm regions, the outputs of the Gabor filter are given negative values. Therefore, the output $G_{gbr}(i, j)$ is determined to have the minimum value of the filtered values.

$$G_{gbr}(i, j) = \min\{G_a(i, j), G_b(i, j), G_c(i, j),\} \quad (8)$$

Figure 8 shows examples of a microaneurysm enhanced by three filters. The shape index is the strongest enhancer of these filters, but this filter wrongly enhanced a narrow vessel. By combining the outputs of three filters, the microaneurysm is enhanced effectively.

Microaneurysm Detection Based on DCNN

Microaneurysms were detected using the DCNN. It scans the entire image and performs microaneurysm prediction for every pixel, based on the image patch. One DCNN was input with an enhanced image patch, and another DCNN was input with the green channel image of the color image patch. The size of the patch was 21 × 21 pixels.

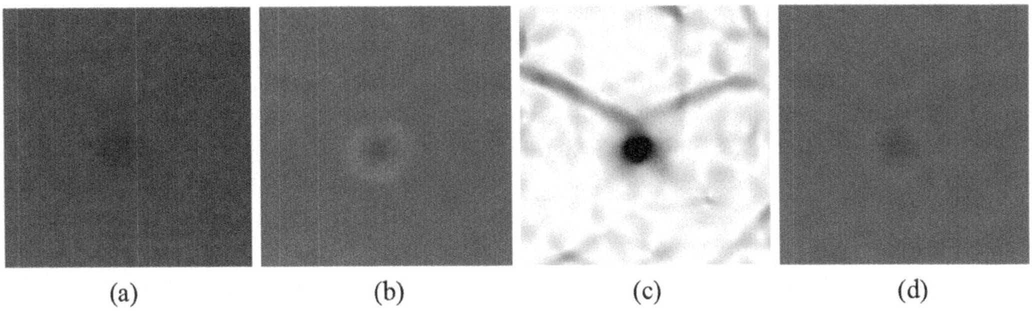

Fig. 8 Examples of enhanced microaneurysm. The microaneurysm is located in the center of image. (**a**) Original image. (**b**) Double-ring filter. (**c**) Shape index. (**d**) Gabor filter

GoogLeNet (Inception-v1) [30] was used for both DCNNs. GoogLeNet won in the ILSVRC2014 (ImageNet Large Scale Visual Recognition Competition), at which time the inception module was first introduced. The key of the inception module is to structure multiple convolutions with multiple filters and pooling layers simultaneously in parallel within the same layer (the inception layer). For example, the input obtained from the previous layer goes through 1×1, 3×3, and 5×5 convolutions, as well as max pooling simultaneously and is concatenated together as output. Therefore, users do not have to think about which filter size should be used for each layer.

Two hundred forty-six microaneurysms from the 83 microaneurysms in DIARETDB1 and the 163 microaneurysms in the Retinopathy Online Challenge database were used for DCNN training. Because the number of microaneurysms was too low, the microaneurysm patches were augmented from 246 to 2460 microaneurysm patches using 9 kinds of image processing, which were horizontal inversion, vertical inversion, horizontal–vertical inversion, moving the microaneurysm to 4 corners, smoothing, and gamma correction. In a preliminary examination, 23 kinds of image processing, which were horizontal inversion, horizontal–vertical inversion, moving the microaneurysm in 8 directions (4 corners, upper, lower, right, and left), 2 kinds of gamma correction, and 11 kinds of rotation (30, 60, 90, ..., 330 degrees), were tested as data augmentation techniques. As a result, using 9 kinds of data augmentation was shown to be better than using 23 kinds of augmentation. Therefore, we used 2460 normal patches and 2460 augmented microaneurysm patches to train the DCNN.

Reducing the Number of False Positives

The candidate microaneurysm regions included so many false positives that the candidates were classified into microaneurysms and false positives using three-layer neural network with 48 features [38]. The neural network parameters, which are objective functions, included the number of units in the hidden layer, weight decay, upper number of cycles, random number seed, and each weights of units, were set by a grid search. Note that the candidates with clear outlier feature values were removed before being input to the neural network.

The 48 features consisted of shape, pixel intensity, location, and statistical texture features. The shape features were area, circularity, length-to-width ratio, and the number of labels in a region of interest, and they were calculated in the binarized image. The color intensity features were mean value in RGB, max-min pixel value, contrast, double-ring filter values, and values of another double-ring filter [8], and they were each calculated in the red, green, and blue channel images. The similarity of blood vessels was calculated in the green channel image as a gray intensity feature. The nearest distance from blood vessels, and the distance to the ordinate and abscissa from the optic disc were calculated in the binarized image as location features. Finally, 12 kinds of Haralick features [44] based on the co-occurrence matrix, 2 kinds of Weszka features [45] based on the gray level difference statistics, and each of 5 kinds of Galloway features [46] of the 0 and 90 deg run-length matrices were calculated in the green channel image as statistical texture features.

Examination

Using test images from the DIARETDB1 database, the method based on the DCNN was tested using free-response ROC (FROC) curves, as shown in Fig. 9. The results of the three filter-based methods [39] and the FCN-based method [41] are also shown in Fig. 9. The overall performance of the DCNN-based method was superior to the other two methods. In this result, the previous false positive reduction method [39] was applied to the proposed method with no change. Therefore, the false positive reduction method should improve the method.

Figure 10 shows an example of results using the DCNN with the reduction in false positives. The DCNN detected many false positives, but the false positive reduction step removed all false positives in this case.

The performance results of the proposed method showed that the sensitivity was 90% for 8.0 false positives per image. One of the problems of the previous method [39] was the existence of too many rules. Several parameters, such as the threshold values of the double-ring filter, Gabor filter, and shape index, as well as parameters for combining the microaneurysm detectors, were all experimentally determined from test results. In the DCNN-based method, although it was necessary to determine the parameter enhancement filter, the other parameters could be automatically learned. The method could be further improved by optimizing the network architecture and adding a post-processing method.

Conclusion

This chapter describes automated blood vessel extraction and automated microaneurysm detection using a DCNN. Although the study of blood vessel extraction using FCN is popular, classic DCNNs like LeNet, AlexNet, etc. are also effective for blood vessel extraction. Moreover, AlexNet can perform classification of arteries and veins without over-fitting. It is difficult to detect microaneurysms because they are very small and have low contrast. However, the DCNN detects them by improving the images input into the DCNN. This chapter showed that the DCNN works effectively for retinal image analysis. The

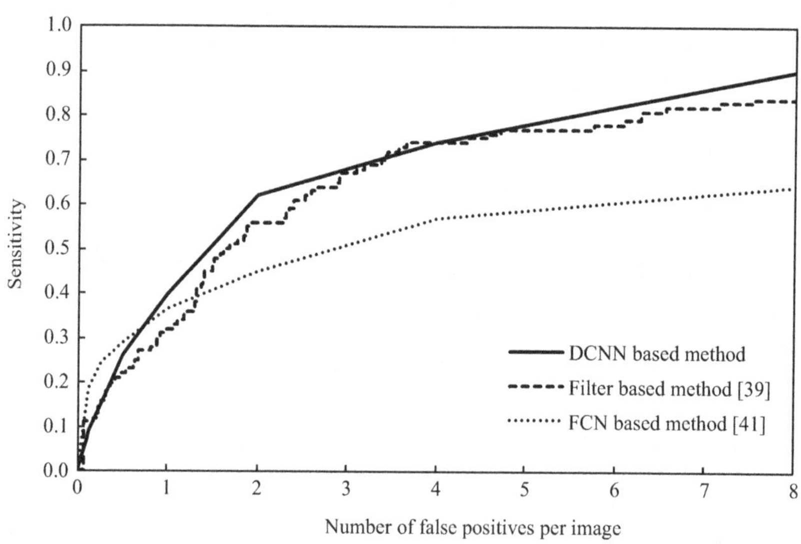

Fig. 9 FROC curves for the 61 test images from the DIARETDB1 database

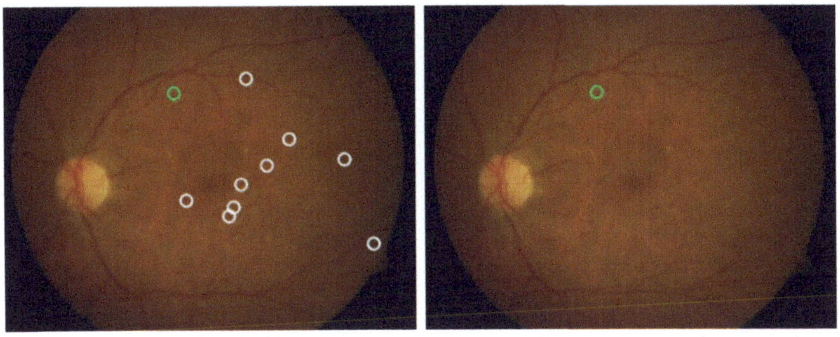

Fig. 10 Example of microaneurysm detection. Left image shows a result of microaneurysm detection using DCNN without false positives reduction. Right image shows a final result. Green circle shows true microaneurysm, and white circles show false positives

proposal of a useful method based on the DCNN is hoped for in the future.

References

1. Scheie HG (1953) Evaluation of ophthalmoscopic changes of hypertension and arteriolar sclerosis. Arch Ophthalmol 49(2):117–138
2. Wong TY, Mitchell P (2004) Hypertensive retinopathy. N Engl J Med 351(22):2310–2317
3. Soares JV, Leandro JJ, Cesar Júnior RM, Jelinek HF, Cree MJ (2006) Retinal vessel segmentation using the 2-D Gabor wavelet and supervised classification. IEEE Trans Med Imaging 25(9):1214–1222
4. Rangayyan RM, Ayres FJ, Oloumi F, Oloumi F, Eshghzadeh-Zanjani P (2008) Detection of blood vessels in the retina with multiscale Gabor filters. J Electron Imaging 17(2):023018
5. Ricci E, Perfetti R (2007) Retinal blood vessel segmentation using line operator and support vector classification. IEEE Trans Med Imaging 26(10):1357–1365
6. Staal J, Abràmoff MD, Niemeijer M, Viergever MA, van Ginneken B (2004) Ridge based vessel segmentation in color images of the retina. IEEE Trans Med Imaging 23(4):501–509
7. Akatsuka Y, Oost E, Shimizu A, Kobatake A, Furukawa D, Katayama A (2010) Vessel segmentation in eye fundus images using ... eye fundus image based on ensemble learning and a classifier cascade. IEICE Tech Rep 109(407):143–148
8. Hatanaka Y, Nakagawa T, Hayashi Y, Aoyama A, Zhou X, Hara T, Fujita H, Mizukusa Y, Fujita A, Kakogawa M (2005) Automated detection algorithm for arteriolar narrowing on fundus images. In: Proceedings of the 27th Annual International Conference of the IEEE-EMBS 2005, pp 286–289
9. Nakagawa T, Hayashi Y, Hatanaka Y, Aoyama A, Mizukusa Y, Fujita A, Kakogawa M, Hara T, Fujita H, Yamamoto T (2006) Recognition of optic nerve head using blood-vessel-erased image and its application to production of simulated stereogram in computer-aided diagnosis system for retinal images. IEICE Trans Inf Syst J89-D(11):2491–2501
10. Iwase T, Muramatsu C, Hatanaka Y, Zhou X, Hara T, Fujita H (2009) Automated detection and classification of major arteries and veins for arteriolar narrowing analysis on retinal fundus images. IEICE Tech Rep 109(407):189–193
11. Hatanaka Y, Samo K, Ogohara K, Sunayama W, Muramatsu C, Okumura S, Fujita H (2017) Automated blood vessel extraction based on high-order local autocorrelation features on retinal images. In: VipIMAGE2017, vol 27, pp 803–810
12. Fu H, Yanwu X, Wong D, Jiang L (2016) Retinal vessel segmentation via deep learning network and fully-connected conditional random fields. In: Proceedings of the IEEE 13th International Symposium on Biomedical Imaging, pp 698–701
13. Martina M, Pavle P, Sven L (2015) Retinal vessel segmentation using deep neural networks. In: Proceedings of the VISAPP, vol 1, pp 577–581
14. Dasgupta A, Singh S (2017) A fully convolutional neural network based structured prediction approach towards the retinal vessel segmentation. In: Proceedings of the ISBI, pp 248–251
15. Ikawa H, Hatanaka Y, Sunayama W, Ogohara K, Muramatsu C, Fujita H (2019) Arteriovenous classification method using convolutional neural network for early detection of retinal vascular lesion. In: Proceedings of the SPIE 11050 International Forum on Medical Imaging in Asia 2019, p 110501M
16. Foracchia M, Grisan E, Ruggeri A (2005) Luminosity and contrast normalization in retinal images. Med Image Anal 9(3):179–190
17. Le Q, Karpenko A, Ngiam J, Ng A (2011) Ica with reconstruction cost for efficient overcomplete feature learning. Adv Neural Inf Process Syst:1017–1025
18. Friedman J (1987) Exploratory projection pursuit. J Am Stat Assoc 82(397):249–266

19. Bell A, Sejnowski T (1997) The "Independent Components" of natural scenes are edge filters. Vis Res 37(23):3327–3338
20. LeCun Y, Bottou L, Bengio Y, Haffner P (1998) Gradient-based learning applied to document recognition. In: Proceedings of the IEEE, pp 1–46
21. Krizhevsky A, Sutskever I, Hinton GE (2012) ImageNet classification with deep convolutional neural networks. Proc Neural Inf Process Syst 2012:1–9
22. Niemeijer M, Xu X, Dumitrescu AV, Gupta P, Ginneken B, Folk JC, Abrámoff M (2011) Automated measurement of the arteriolar-to-venular width ratio in digital color fundus photographs. IEEE Trans Med Imaging 30(11):1941–1950
23. Dashtbozorg B, Mendonca AM, Penas S, Campilho A (2014) RetinaCAD, a system for the assessment of retinal vascular changes. In: Proceedings of the 36th Annual International Conference of the IEEE Engineering in Medicine and Biology Society, pp 6328–6331
24. Vázquez SG, Cancela B, Barreira N, Penedo MG, Rodríguez-Blanco M, Pena Sei-jo M, Coll de Tuero G, Barceló MA, Saez M (2013) Improving retinal artery and vein classification by means of a minimal path approach. Mach Vis Appl 24(5): 919–930
25. Muramatsu C, Hatanaka Y, Iwase T, Hara T, Fujita H (2011) Automated selection of major arteries and veins for measurement of arteriolar-to-venular diameter ratio on retinal fundus images. Comput Med Imaging Graph 35(6):472–480
26. Hatanaka Y, Tachiki H, Okumura S, Ogohara K, Muramatsu C, Fujita H (2017) Automated independent extraction of major arteries and veins on retinal images. Med Imaging Inf Sci 34(3):136–140
27. Meyer MI, Galdran A, Costa P, Mendonça AM, Campilho A (2018) Deep convolutional artery/vein classification of retinal vessels. In: ICIAR 2018 Lecture Notes in Computer Science, vol 10882, pp 622–630
28. Girard F, Kavalec C, Cheriet F (2019) Joint segmentation and classification of retinal arteries/veins from fundus images. Artif Intell Med 94:96–109
29. Hu Q, Abràmoff MD, Garvin MK (2013) Automated separation of binary overlapping trees in low-contrast color retinal images. In: Medical Image Computing and Computer-Assisted Intervention (MICCAI 2013), pp 436–443
30. Szegedy C, Liu W, Jia Y, Sermanet P, Reed S, Anguelov D, Erhan D, Vanhoucke V, Rabinovich A (2015) Going deeper with convolutions. In: IEEE Conference on Computer Vision and Pattern Recognition, pp 1–9
31. Knudtson MD, Lee KE, Hubbard LD, Wong TY, Klein R, Klein BEK (2003) Revised formulas for summarizing retinal vessel diameters. Curr Eye Res 27(3):143–149
32. Cheung N, Mitchell P, Wong TY (2010) Diabetic retinopathy. Lancet 376(9735):124–136
33. Niemeijer M, van Ginnerken B, Cree MJ, Mizutani A, Quellec G, Sanchez CI, Zhang B, Hornero R, Lamard M, Muramatsu C, Wu X, Cazuquel G, You J, Mayo A, Li Q, Hatanaka Y, Cochener B, Roux C, Karray F, Garcia M, Fujita H, Abramoff MD (2010) Retinopathy Online Challenge: automatic detection of microaneurysms in digital color fundus photographs. IEEE Trans Med Imaging 29(1):185–195
34. Adal KM, Sidibe D, Ail S, Chaum E, Karnowski TP, Mériaudeau F (2014) Automated detection of microaneurysms using scale-adapted blood analysis and semi-supervised learning. Comput Methods Programs Biomed 114(1):1–10
35. Antal B, Hajdu A (2012) An ensemble-based system for microaneurysm detection and diabetic retinopathy grading. IEEE Trans Biomed Eng 59(6):1720–1726
36. Seoud L, Hurtut T, Chelbi J, Cheriet F, Langlois JM (2016) Red lesion detection using dynamic shape features for diabetic retinopathy screening. IEEE Trans Med Imaging 35(4):1116–1126
37. Dai B, Wu X, Bu W (2016) Retinal microaneurysms detection using gradient vector analysis and class imbalance classification. PLoS One 11(8):0161556
38. Inoue T, Hatanaka Y, Okumura S, Ogohara K, Muramatsu C, Fujita H (2015) Automatic microaneurysm detection in retinal fundus images by density gradient vector concentration. J Inst Image Electron Eng Jpn 44(1):58–66
39. Hatanaka Y, Inoue T, Ogohara K, Okumura S, Muramatsu C, Fujita H (2018) Automated microaneurysm detection in retinal fundus images based on the combination of three detectors. J Med Imaging Health Inform 8(5):1103–1112
40. Haaloi M (2016) Improved microaneurysm detection using deep neural networks. arXiv:1505.04424v2
41. Chudzik P, Majumdar S, Caliva F, Al-Diri B, Hunter A (2018) Microaneurysm detection using fully convolutional neural networks. Comp Methods Programs Biomed 158:185–192
42. Miyashita M, Hatanaka Y, Ogohara K, Muramatsu C, Sunayama W, Fujita H (2018) Automatic detection of microaneurysms in retinal image by using convolutional neural network. Med Imaging Technol 36(4):189–195
43. Kauppi T, Kalesnykiene V, Kamarainen JK, Lensu L, Sorri I, Raninen A, Voutilainen R, Uusitalo H, Kälviäinen H, Pietilä J (2007) Diaretdb1 diabetic retinopathy database and evaluation protocol. In: Proceedings of the 11th Conference on Medical Image Understanding and Analysis 2007, pp 61–65
44. Haralick RM, Shanmugam K, Dinstein I (1973) Textural features for image classification. IEEE Trans Syst Man Cybern 3(6):610–621
45. Weszka JS, Dyer CR, Rosenfeld A (1976) A comparative study of texture measures for terrain classification. IEEE Trans Syst Man Cybern 6(4):269–285
46. Galloway MM (1975) Texture analysis using gray level run lengths. Comput Graph Image Process 4(2):172–179

Diagnosis of Glaucoma on Retinal Fundus Images Using Deep Learning: Detection of Nerve Fiber Layer Defect and Optic Disc Analysis

Chisako Muramatsu

Abstract

Early detection of glaucoma is important to slow down progression of the disease and to prevent total vision loss. Retinal fundus photography is frequently obtained for various eye disease diagnosis and record and is a suitable screening exam for its simplicity and low cost. However, the number of ophthalmologists who are specialized in glaucoma diagnosis is limited. We have been studying automated schemes for detection of nerve fiber layer defects and analysis of optic disc deformation, two major signs of glaucoma, in assisting ophthalmologists' accurate and efficient diagnosis. In this chapter, our recent progress in computerized methods is discussed.

Keywords

Retinal fundus images · Glaucoma · Nerve fiber layer defect · Cup-to-disc ratio · Detection · Segmentation · Classification · Convolutional neural network

C. Muramatsu (✉)
Faculty of Data Science, Shiga University, Hikone, Japan
e-mail: chisako-muramatsu@biwako.shiga-u.ac.jp

Introduction

Glaucoma is the second leading cause of vision loss in the world, and the number of affected patients is expected to increase in the aging society up to about 80 million by 2020 [1]. Damage to retinal nerves cannot be recovered; therefore, the early detection and appropriate treatment are important to cease or slow down the progression for preventing total blindness. However, due to slow progressive nature, patients stay asymptomatic during the early stages, resulting in delay of diagnosis. In fact, a population-based prevalence survey of glaucoma in Tajimi, Japan showed that 93% of the primary open-angle glaucoma patients were previously undiagnosed, suggesting the need for an effective screening paradigm [2].

Retinal examination using fundus photographs is rather simple and relatively inexpensive way for screening glaucoma. It is a common exam performed at general ophthalmologic visits for diagnosis, clinical record, and treatment follow-up. Some of the early signs of glaucoma found in fundus photographs include retinal nerve fiber layer defect (NFLD) and optic disc deformation. Generally NFLD development precedes an occurrence of morphologic change in optic disc and visual field defects. NFLD appears as a strip- or

G. Lee, H. Fujita (eds.), *Deep Learning in Medical Image Analysis*, Advances in Experimental Medicine and Biology 1213,
https://doi.org/10.1007/978-3-030-33128-3_8

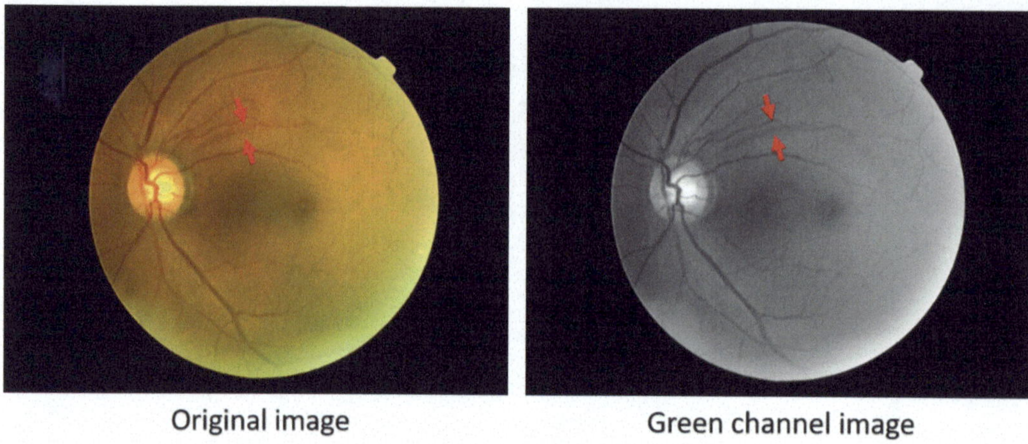

Fig. 1 Retinal fundus photograph with NFLD. NFLD is observed between arrows

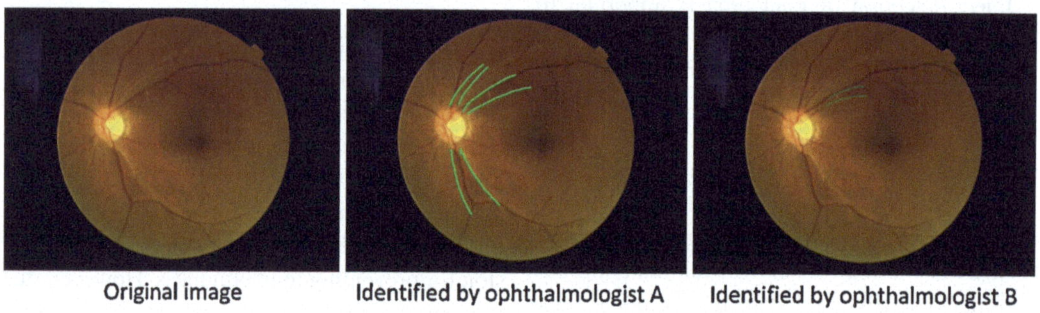

Fig. 2 Variation in identification of NFLD regions by two ophthalmologists. NFLD regions were marked by two lines. Ophthalmologist A identified three NFLDs, whereas ophthalmologist B marked only one NFLD

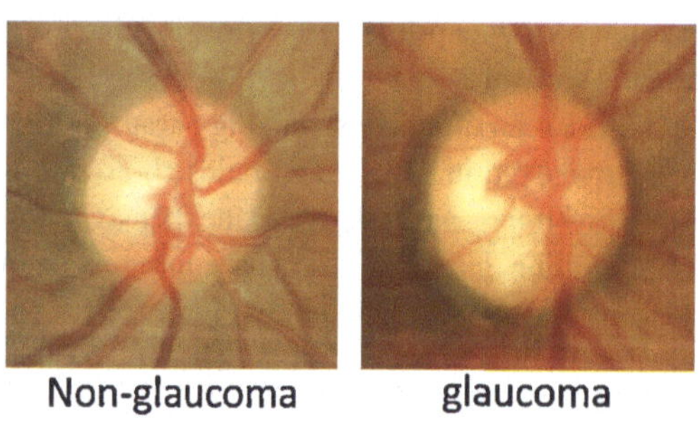

Fig. 3 Non-glaucomatous and glaucomatous disc images. Cup region (generally more bright region inside the optic disc) is enlarged in glaucomatous eye

wedge-shaped region darker than the surrounding retina as shown in Fig. 1. Some early, localized defects are very subtle, and assessment by experienced ophthalmologists could even differ (Fig. 2).

As the disease progresses, optic cup is enlarged. Figure 3 shows the non-glaucomatous and glaucomatous discs. Usually the cupping or rim thinning is more apparent in upper or lower temporal sides of optic disc in early stages. Therefore, the ratio of vertical diameters of cup and disc is considered as a diagnostic index. However, it is hardly measured during the clinical practice. Even if it is measured, it is prone to intra-reader and inter-reader variations [3, 4], as shown in Fig. 4. Furthermore, the number of ophthalmologists specialized in glaucoma diagnosis is limited. Computerized analysis of fundus images could assist ophthalmologists in accurate, efficient, and consistent diagnosis by providing suspected condition, location, and quantitative information. We have been studying the computerized methods to detect NFLDs and to analyze optic disc for assisting early diagnosis. In this chapter, our recent progress in the computerized methods is introduced.

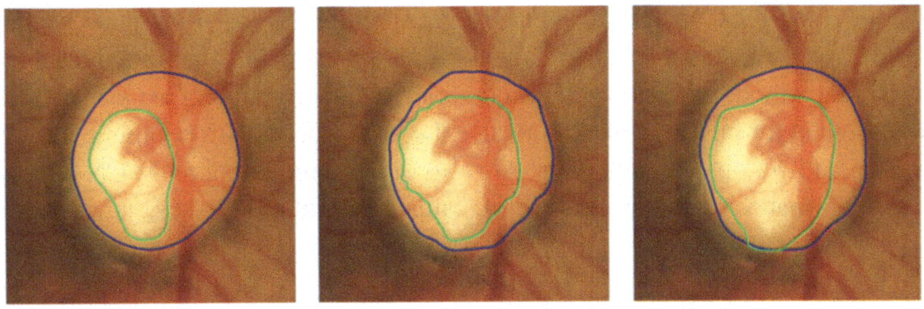

Fig. 4 Variation in optic disc and cup sketches between three ophthalmologists. Blue and green lines specify disc and cup regions, respectively

Related Works

There have been many studies on computer analysis of fundus images for diagnosis of glaucoma. Conventional analysis on NFL includes studies for estimating NFL thickness [5] and for detecting NFLDs [6, 7] based on image features such as texture features using classifiers such as a support vector machine and random forest. Other studies proposed various methods to segment cup and disc regions for calculation of cup-to-disc ratio [8–11].

Recently these methods have been replaced by ones using deep learning. Panda et al. proposed NFLD detection method based on intensity profile analysis [12]. After the NFLD boundary candidates were found, they used a convolutional neural network to differentiate between true and false boundary patches. Medeiros et al. trained the ResNet [13] to estimate average NFL thickness from retinal fundus images [14]. The input data were the photographs of the disc region, and the teacher data were the measurement from the optical coherent tomography (OCT) of the corresponding patients. They obtained the correlation coefficient of 0.83 between the predicted values and OCT measurements.

Since the first-reader or prescreening type of diagnostic system for diabetic retinopathy has been approved by the Food and Drug Administration, Christopher et al. investigated the possibility of such a system for glaucoma [15]. They compared the performance of the popular networks, namely, VGG16 [16], inception-v3 [17], and ResNet50 [13] for classification of glaucomatous optic neuropathy (GON) and non-GON disc images. They obtained the area under the receiver operating characteristic curve (AUC) of 0.91 using ResNet. The heatmap analysis indicated the importance of inferior and superior parts of neuroretinal rim in the classification. Shibata et al. compared their system, also using the ResNet, with residents [18]. The classification AUC for 110 test cases by the network was higher than those by the three residents. The network showed the high performance for classification of challenging cases between the myopic glaucoma and myopic normal patients.

Fu et al. proposed an ensemble network which consists of four networks that analyze local and global information [19]. The two networks based on ResNet architecture take disc region images and polar transformed disc region images to classify between glaucoma and normal. The third network with the u-net [20] architecture takes the whole fundus images to localize disc region and to classify between glaucoma and normal with the branch layers. The fourth network based on the ResNet takes the whole images for classification directly. They reported the superior performance of their ensemble network compared with the single ones.

NFLD Detection

Background

As mentioned previously, NFLD is one of the earliest signs of glaucoma that can be found on fundus photographs. We have previously proposed a computerized method for detection of NFLDs using a modified polar transform and the Gabor filter [6, 21]. In the method, various handcrafted features were designed for false positive removal using a classifier and rule-based approach. Although a good performance was obtained, use of the empirically optimized rules was a concern. For further improving the detection performance and achieving robustness, we propose NFLD detection systems using convolutional neural networks (CNNs), and discuss two types of networks.

Proposed Method

Segmentation Network

The first network is based on the fully convolutional network (FCN) proposed by Long et al. [22]. The original network is trained in 3 stages: FCN-32s consists of 16 convolutional layers and 4 pooling layers in downsampling path and an upsampling layer with stride 32 in a single step; FCN-16s, which is fine-tuned from the FCN-32s, carries out upsampling in two steps and combines the last layer and 4th pooling layer with stride 16 to output prediction; and FCN-8s, which is fine-tuned from the FCN-16s, carries out upsampling in three steps and combines the last layer and 3rd pooling layer with stride 8 to output predictions. The network weights were initialized with the VGG net trained with the PASCAL [23] dataset.

Figure 5 shows the segmentation network used in this study. The input is the fundus image clipped to 576 × 576 pixels and the teacher is the binary image specifying the NFLD and background regions. Because the NFLD regions are relatively narrow, we removed the pool4 and conv5 layers from the original network so that the images are downsampled to 1/16. Two skip connections are added from pool2 and pool3 layers. The initial weights were transferred from the PASCAL-trained VGG16 except for the

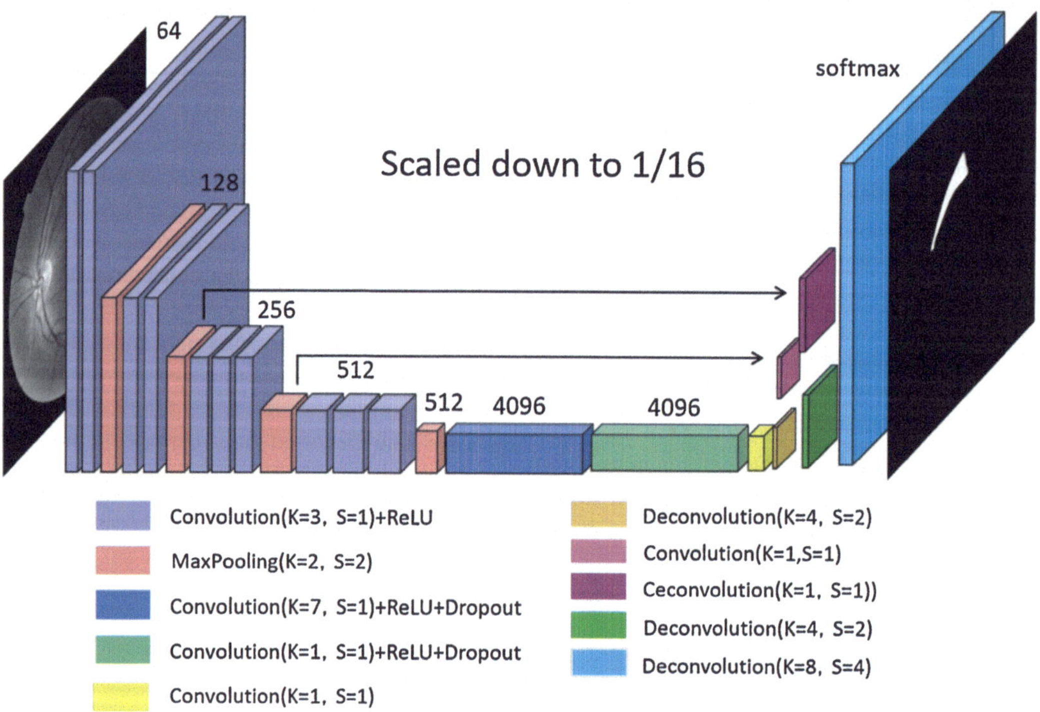

Fig. 5 Segmentation network architecture used in this study (modified from [22]). Above the boxes are the number of filters. Arrows specify skip connections

removed layers, and the network was trained at one step.

Detection Network

In our conventional method, the gold standard was the NFLD regions specified by an expert ophthalmologist rather than locations (e.g., center pixels), and hence our output. It seemed natural to use the segmentation network to identify NFLD regions. However, the goal of the diagnostic system is to suggest presence of possible NFLDs and their approximate location. For this purpose, pixel-by-pixel region determination is unnecessary and box-based object detection is appropriate. Therefore, we investigated the use of a detection network and compared the results with those by the segmentation network.

The network is based on the single shot multibox detector (SSD) proposed by Liu et al. [24]. It consists of the VGG layers up to conv5 layers and several downsampling convolutional layers to extract features and to detect target objects at multiple scales. The pretrained weights from VGG net are also used in the SSD. Because the NFLD regions generally occupy only small parts of fundus images, we removed last two convolutional layers in this study. The network architecture used is shown in Fig. 6.

Combined Method

Using the segmentation network, the probability of being a part of NFLD regions can be estimated for each pixel. Such information can be incorporated as a prior knowledge in the detection network for improving performance. To test the hypothesis, the detection network was trained with the original image combined with the probability-weighted image. Both the segmentation and detection networks take color images as input to utilize the pretrained VGG network. However, the blue component of fundus images, in general, has few information while NFLDs are most prominent in the green channel, as shown in Fig. 7. Hence in this study, we employed the green component for all three input planes. In the combined method, one plane is replaced by a probability-weighted image as shown in Fig. 8.

Dataset

The images used in this study were obtained from the database collected as a part of the Eye Health Care Project in Tajimi, Japan [2], which

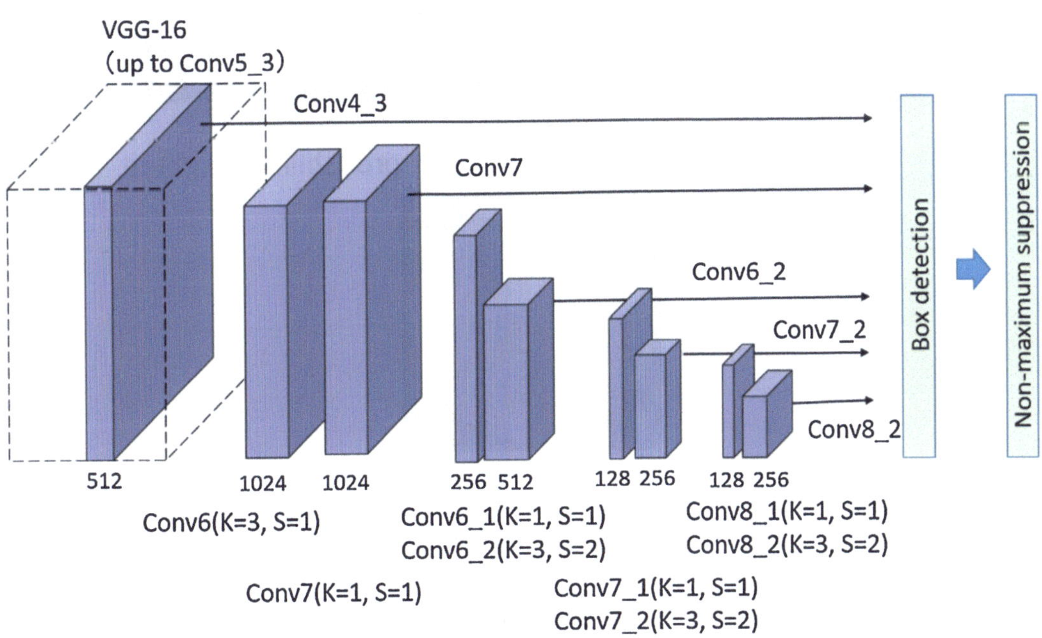

Fig. 6 Detection network architecture used in this study (modified from [24])

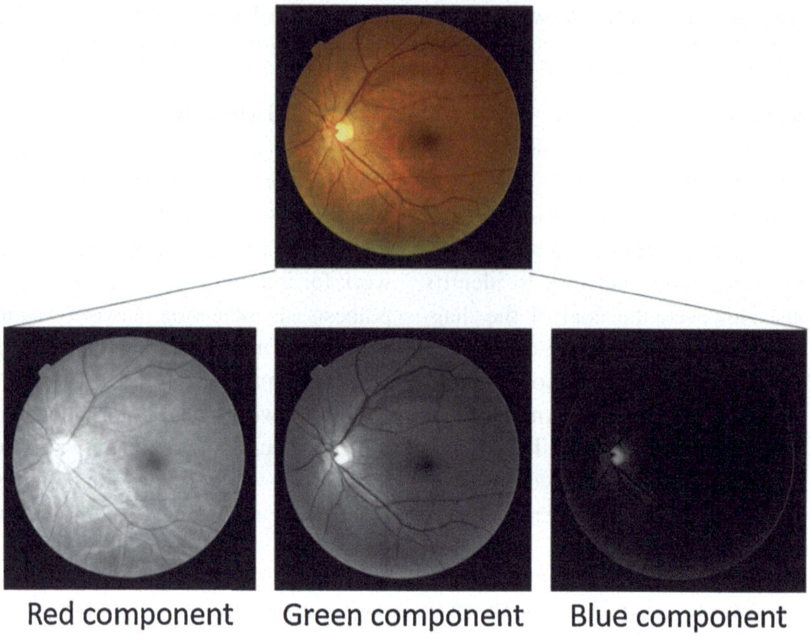

Fig. 7 Color components of a fundus image. Top is the original image, and bottom three images are three color components. Green channel has the most information regarding NFLDs

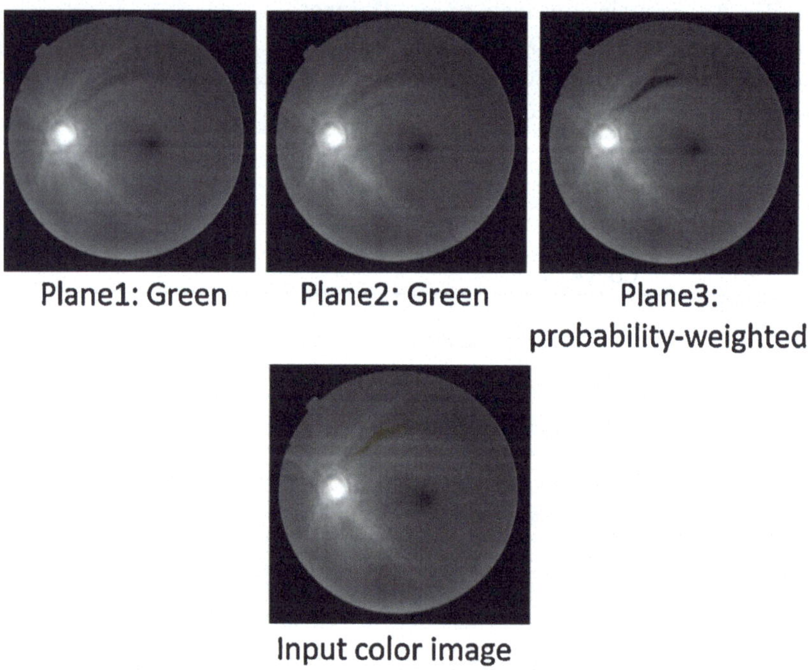

Fig. 8 Input image for the combined method using the probability-weighted image. Red and blue channels are replaced by green and probability image, respectively, for a combined input image

includes the clinical data and eye examination results such as visual field test, intraocular pressure, and retinal fundus images. The cases used in the randomized study [2] were not included in this study. The fundus photographs were obtained with an IMAGEnet digital fundus camera system (TRC-NW6S, Topcon, Tokyo, Japan). The original image size was 768 × 576 pixels, which was cropped to 576 × 576 pixels before input to the networks.

The training set includes 81 images of left eyes that were identified as having NFLDs on the basis of the image finding report. These images were retrospectively reviewed by two ophthalmologists independently, and 99 regions identified as NFLDs by both ophthalmologists were considered as the gold standard in this study.

Test dataset including images of the right eyes was collected on the basis of the image finding report. None of the images was of the same patients in the training set. The dataset includes 130 cases with NFLDs and 131 age- and gender-matched cases without NFLDs; one duplicate NFLD case was excluded. One ophthalmologist reviewed the images retrospectively and identified 203 NFLD regions without any clinical information in the same way as the training set.

Preprocessing

Localized NFLD regions are depicted as dark strip-like region as shown in Fig. 1. Retinal vessels, also depicted as dark tubes, can often become sources of false positives. As in the conventional study [6], we preprocessed the images to remove retinal vessels before input to the networks. The detailed procedure for creating vessel erased images has been described elsewhere [25]. Briefly, the retinal vessel regions are detected using black tophat filter, and the detected regions are interpolated by the weighted average of the surrounding pixels. The images before and after the vessel erasing procedure can be found in Figs. 7 and 8, respectively.

Evaluation

The performance of the proposed networks was evaluated by the free response receiver operating characteristic (FROC) analysis. In our previous study, the NFLD regions were marked by the ophthalmologists with two enclosing lines, and the gold standard masks were created by filling in the two lines. Using the detection network, on the other hand, the teacher data and the output were the bounding boxes. For fair comparison, the NFLD masks and the outputs from the segmentation network were bounded by the smallest rectangles, and these boxes were used as the gold standard and the results from the segmentation network, respectively. If more than a half of the output box overlapped with the gold standard box, it is counted as the true detection.

Results

Detection results using the segmentation network are shown in Fig. 9. In these cases, the most of the NFLD regions are correctly detected with a few small false positives. When the detection threshold is raised to remove some of these false positives, true positives are also removed. On the other hand, lowering the threshold above certain point would not improve sensitivity as some detected regions would be connected. Figure 10 shows the results by the detection network. At a low threshold, the high sensitivity is obtained albeit a large number of false positives. It can be seen that many overlapped boxes with different sizes are detected by the network.

For keeping the high sensitivity and decreasing the false positives, the use of the segmentation output as input data for the detection network was proposed. The FROC curves for the three methods are shown in Fig. 11. By the combined method, the sensitivity is improved with the small number of false positives. The sensitivity is 80% at the 0.59 false positives per image.

Discussion

Fair results were obtained using both networks with their suitable loss functions. The output probabilities from the segmentation network tend to be bipolarized; the pixels are marked as a part of the NFLD regions with very high

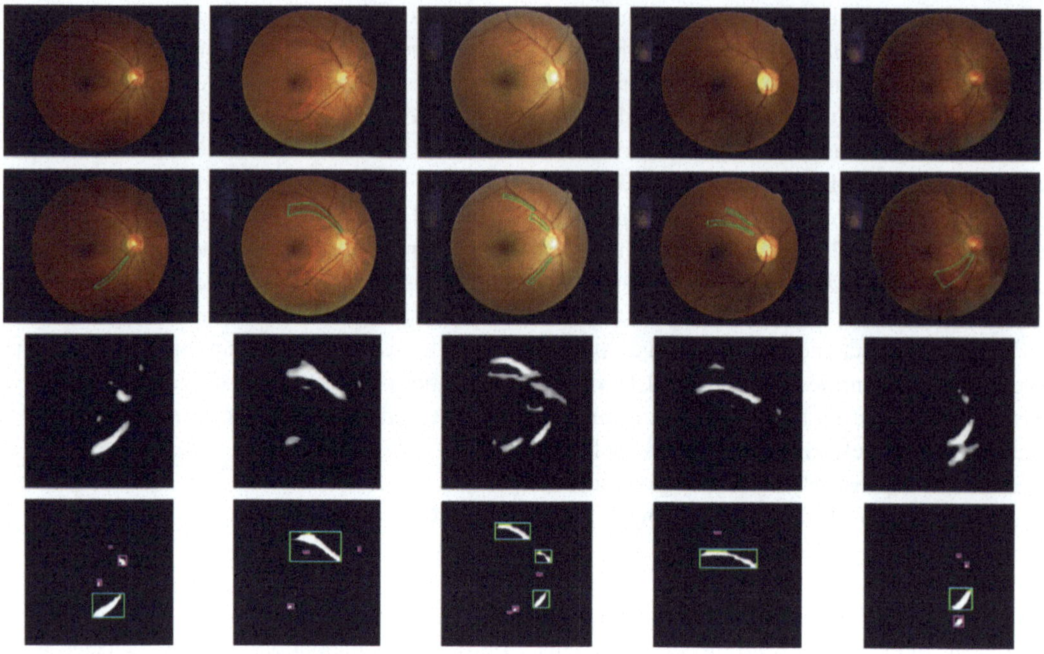

Fig. 9 Results by the segmentation network. From the first to the bottom rows: original images, the gold standard regions, probability output from the network, and the detection results (green box: true detection, pink box: false positive detection)

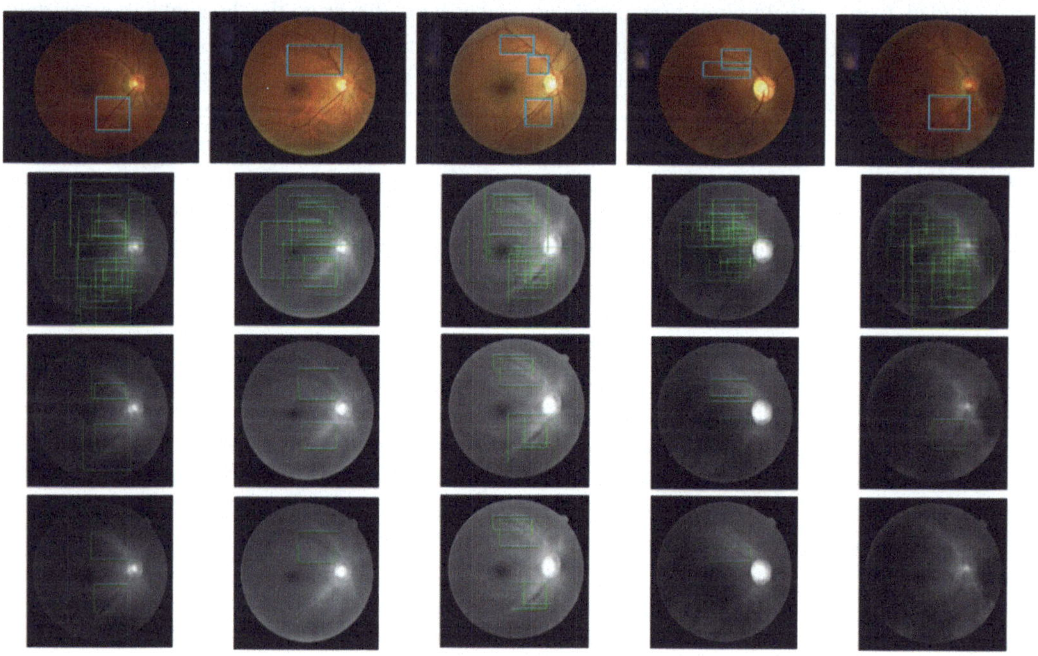

Fig. 10 Results of the detection network. From the first to the bottom rows: the gold standard boxes, and the detection results at low, medium, and high thresholds

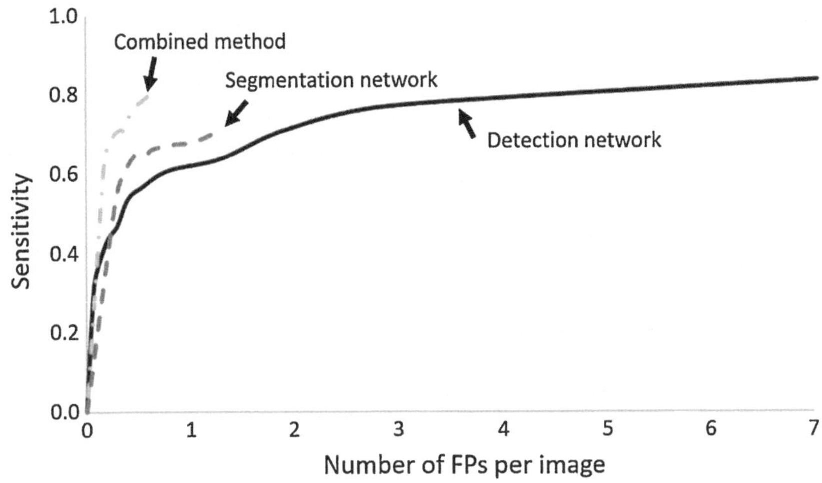

Fig. 11 FROC curves of the proposed methods

confidence or very low confidence. Therefore, at a certain threshold (generally very low), a large area is detected as one large NFLD. By the detection network, multi-size boxes are considered; therefore, the high sensitivity can be obtained but with a large number of false positive boxes that are overlapped. In this study, the probability map from the segmentation network was employed as an input data for the detection network because the precise segmentation is not our goal in this study. Using the combined method, the sensitivity was improved at low false positive rate. Although result was promising, further investigation is needed for post processing methods and optimization of the networks. Whether such system can help physicians in fundus image diagnosis and how to effectively display the results are our future studies.

Optic Disc Analysis

Background

Optic disc deformation is another major finding of glaucoma. When the retinal nerves are damaged, rim thinning occurs. In our previous study, we proposed computerized methods for quantitative measurement of cup-to-disc ratio (CDR) on stereo fundus photographs [26] and plain photographs [27]. With the stereo images, depth images were created, and the disc and cup regions were segmented using the active contour method and dynamic programing. For the plain photographs, intensity profile was used to determine the cup boundary. The computed CDRs were effective in distinguishing between glaucoma and non-glaucoma eyes; however, it was difficult to evaluate the accuracy of CDR, because there is no true value and the measurements by ophthalmologists vary. In this study, we investigated the classification method that directly estimates the glaucoma probability from the optic disc images using CNN.

Methods

The network used for the classification is the AlexNet model [28]. First, a region of interest with 600 × 600 pixels including optic disc is clipped on the basis of the automated disc detection result [26]. We considered three input images for comparison: (1) an original image, (2) a vessel erased image, and (3) a depth information composite image. For (1) and (2), input images are color images, and the vessel erased images were created as described in the previous section. For

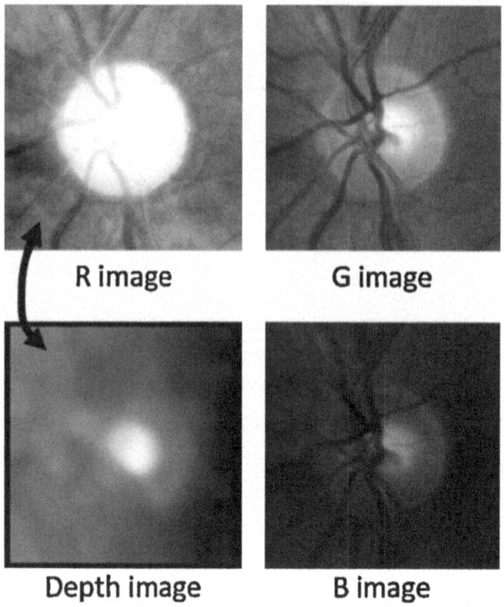

Fig. 12 Depth image and RGB components of the disc images. Red channel was replaced by the depth image

(3), depth images were created using the stereo image pairs as described in [26]. Figure 12 shows an example of depth image. As shown in Fig. 12, generally the red component has little information inside the disc region; therefore, the red plane image was replaced by the depth image, and the composite image was employed as the input. Because of the small dataset, we investigated the effect of data augmentation by rotation (±5 and 10 degrees) and intensity transformation (gamma transformation with $\gamma = 0.8, 0.9, 1.1$, and 1.2). The classification performance was evaluated by ROC analysis.

Dataset

Images used in this study were obtained with a stereo fundus camera (a prototype of WX-1; Kowa Company, Ltd.). A stereo image pair was saved as a single JPEG image with 2144 × 1424 pixels. The images were separated in the middle, and the left image was employed in this study, except for depth calculation. As described earlier, the optic disc location was automatically detected and the ROI was clipped. The images were retrospectively reviewed by two ophthalmologists experienced in glaucoma diagnosis, and the cases in which diagnosis by both reviewers and the result of visual field test agreed were included in this study. The database consists of 148 glaucomatous and 153 non-glaucomatous cases. Only one eye per patient was included.

From both groups, 15 images were selected for testing and the remaining images were employed for training. For reducing the selection bias, this process was repeated for 10 times.

Results

Figure 13 shows the ROC curves. The vessel removal and the depth information were useful, although there was no statistical significant difference in AUCs using the original and vessel erased images. Table 1 shows the AUC values and the p-values against the result of the original images. Data augmentation was also useful; therefore, all three processes were performed. The superior performance with the AUC of 0.92 was obtained.

Discussion

In this study, we investigated the use of the vessel erased images as this process was useful in our previous study for creating depth images. However, in clinical practice, vessel location is important information for glaucoma diagnosis. As the disease progresses and cup region is enlarged, the vessels are shifted to nasal side. Sometimes, vessel may appear from the disc edges (bayonetting), as shown in Fig. 14. Although AUC was slightly improved using the vessel erased images, utility of the vessel information must be examined in the future. The result indicated the proposed method is useful in classification of glaucomatous and non-glaucomatous eyes. Our future study is to provide the location of suspicious findings that most contributed in the classification.

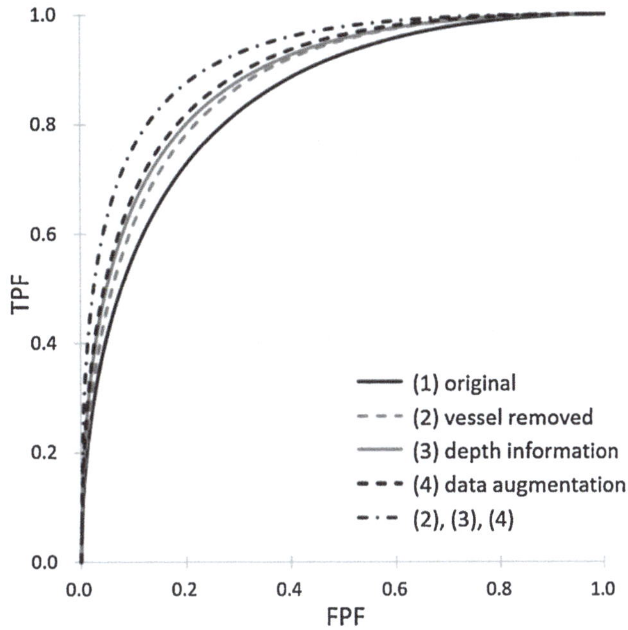

Fig. 13 ROC curves using different input images and data augmentation techniques

Table 1 AUC values and p-values for classification of glaucoma and non-glaucoma images

	(1) Original	(2) Vessel removal	(3) Depth information	(4) Data augmentation	(5) Processes (2), (3) and (4)
AUC	0.846	0.873	0.882	0.891	0.920
p-value	–	0.08	0.03	0.008	0.00001

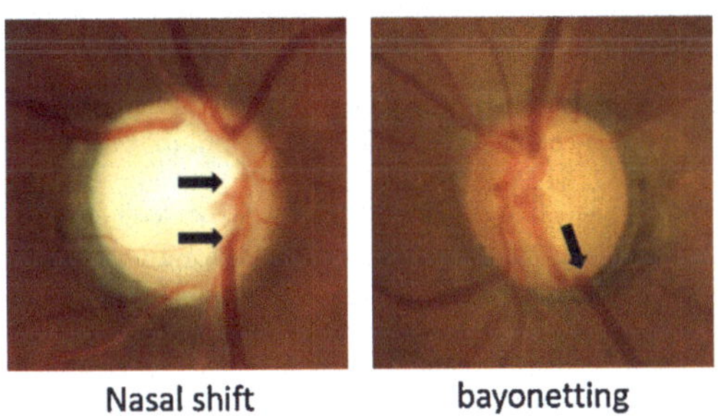

Fig. 14 Cases of nasal shift and bayonetting in which vessel information is useful for glaucoma diagnosis. The arrows specify the corresponding findings

Summary

In this chapter, we introduced our progress in automated analysis of retinal fundus images for glaucoma diagnosis using CNNs. The proposed NFLD detection method is more robust and provided the comparable performance with that of the conventional method using empirical rules. Computation of CDR can be useful; however, the algorithm is trained on the suboptimal gold standard which includes intra- and inter-reader variation. Using the disc images, classification performance was high. It is possible to provide automatically learned characteristics by the network. Our study was preliminary study with rather small datasets. Further investigation is needed with the large database and improved networks.

Acknowledgements The authors thank to Tetsuya Yamamoto, M.D., Akira Sawada, M.D., Kyoko Ishida, M.D., Hiroshi Fujita, Ph.D., and other members of Fujita/Hara Lab at Gifu University for their contribution in this study. This study is partly supported by a research grant by the Mazda Foundation and a Grant-in-Aid for Scientific Research on Innovative Areas (No. 2610805) by MEXT, Japan.

References

1. Quigley H, Broman AT (2006) The number of with glaucoma worldwide in 2010 and 2020. Br J Ophthalmol 90:262–267

2. Iwase A, Suzuki Y, Araie M, Yamamoto T, Abe H, Shirato S et al (2004) The prevalence of primary open-angle glaucoma in Japanese; the Tajimi Study. Ophthalmology 111:1641–1648
3. Verma R, Spaeth GL, Steinmann WC, Katz LJ (1989) Agreement between clinicians and an image analyzer in estimating cup-to-disc ratios. Arch Ophthalmol 107:526–529
4. Tielsch JM, Katz J, Quigley HA, Miller NR, Sommer A (1988) Intraobserver and interobserver agreement in measurement of optic disc characteristics. Ophthalmology 95:350–356
5. Odstrcilik J, Kolar R, Tornow RP, Jan J, Budai A, Mayer M et al (2014) Thickness related textual properties of retinal nerve fiber layer in color fundus images. Comput Med Imaging Graph 38:508–516
6. Muramatsu C, Hayashi Y, Sawada A, Hatanaka Y, Hara T, Yamamoto T et al (2010) Detection of retinal nerve fiber layer defects on retinal fundus images for early diagnosis of glaucoma. J Biomed Optics 15:016021
7. Panda R, Puhan NB, Rao A, Padhy D, Panda G (2018) Automated retinal nerve fiber layer defect detection using fundus imaging in glaucoma. Comput Med Imaging Graph 66:56–65
8. Almazroa A, Burman R, Raahemifar K, Lakshminarayanan V (2015) Optic disc and optic cup segmentation methodologies for glaucoma image detection; a survey. J Ophthalmol. https://doi.org/10.1155/2015/180972
9. Thakur N, Juneja M (2018) Survey on segmentation and classification approaches of optic cup and optic disc for diagnosis of glaucoma. Biomed Sig Proc Control 42:162–189
10. Abramoff MD, Alward WLM, Greenlee EC, Shuba L, Kim CY, Fingert JH et al (2007) Automated segmentation of the optic nerve head from stereo color photographs using physiologically plausible feature detectors. Invest Ophthalmol Vis Sci 48:1665–1673
11. Cheng J, Liu J, Xu Y, Yin F, Wong DWK, Tan NM et al (2013) Superpixel classification based optic disc and optic cup segmentation for glaucoma screening. IEEE Trans Med Imaging 32:1019–1032
12. Panda R, Puhan NB, Rao A, Mandal B, Padhy D, Panda G (2018) Deep convolutional neural network-based patch classification for retinal nerve fiber layer defect detection in early glaucoma. J Med Imaging 5:044003
13. He K, Zhang X, Ren S, Sun J. Deep residual learning for image recognition. arXiv:1512.03385
14. Medeiros FA, Jammal AA, Thompton AC (2019) From machine to machine: an OCT-trained deep learning algorithm for objective quantification of glaucomatous damage in fundus photographs. Ophthalmology 126:513–526
15. Christopher M, Belghith A, Bowd C, Proudfoot JA, Goldbaum MH, Weinreb RN et al (2018) Performance of deep learning architectures and transfer learning for detecting glaucomatous optic neuropathy in fundus photographs. Sci Rep 8:16685
16. Symonyan K, Zisserman A (2015) Very deep convolutional networks for large-scale image recognition. Int Conf Learn Represent
17. Szegedy C, Liu W, Jia Y, Sermanet P, Reed S, Anguelov D et al (2015) Going deeper with convolutions. IEEE Conf Comput Vis Pat Recog
18. Shibata N, Tanito M, Mitsuhashi K, Fujino Y, Matsuura M, Murata H et al (2018) Development of a deep residual learning algorithm to screen for glaucoma from fundus photography. Sci Rep 8:14665
19. Fu H, Cheng J, Xu Y, Zhang C, Wong DWK, Liu J et al (2018) Disc-aware ensemble network for glaucoma screening from fundus images. IEEE Trans Med Imaging 37:2493–2501
20. Ronneberger O, Fischer P, Brox T. U-Net: convolutional networks for biomedical image segmentation. arXiv:1505.04597
21. Muramatsu C, Ishida K, Sawada A, Hatanaka Y, Yamamoto T, Fujita H (2016) Automated detection of retinal nerve fiber layer defects on fundus images: false positive reduction based on vessel likelihood. Proc SPIE Med Imaging 9785:97852L
22. Long J, Shelhamer E, Darrell T. Fully convolutional networks for semantic segmentation. arXiv:1411.4038
23. The PASCAL Visual Object Classes Homepage. http://host.robots.ox.ac.uk/pascal/VOC/. Accessed 5 June 2019
24. Liu W, Anguelov D, Erhan D, Szegedy C, Reed S, Fu CY et al. SSD: single shot multibox detector. arXiv:1512.02325
25. Nakagawa T, Suzuki T, Hayashi Y, Mizukusa Y, Hatanaka Y, Ishida K et al (2008) Quantitative depth analysis of optic nerve head using stero retinal fundus image pair. J Biomed Opt 13:064026
26. Muramatsu C, Nakagawa T, Sawada A, Hatanaka Y, Yamamoto T, Fujita H (2011) Automated determination of cup-to-disc ratio for classification of glaucoma and normal eyes on stero retinal fundus images. J Biomed Optics 16:096009
27. Hatanaka Y, Nagahata Y, Muramatsu C, Okumura S, Ogohara K, Sawada A et al (2014) Improved automated optic cup segmentation based on detection of blood vessel bends in retinal fundus images. Proc IEEE Eng Med Bio Soc:126–129
28. Krizhevsky A, Sutskever I, Hinton GE (2012) ImageNet classification with deep convolutional neural networks. In: Advances in Neural Information Processing System 25 (NIPS2012), vol 1, pp 1097–1105

Part III

Applications: Emerging Opportunities

Automatic Segmentation of Multiple Organs on 3D CT Images by Using Deep Learning Approaches

Xiangrong Zhou

Abstract

This chapter focuses on modern deep learning techniques that are proposed for automatically recognizing and segmenting multiple organ regions on three-dimensional (3D) computed tomography (CT) images. CT images are widely used to visualize 3D anatomical structures composed of multiple organ regions inside the human body in clinical medicine. Automatic recognition and segmentation of multiple organs on CT images is a fundamental processing step of computer-aided diagnosis, surgery, and radiation therapy systems, which aim to achieve precision and personalized medicines. In this chapter, we introduce our recent works on addressing the issue of multiple organ segmentation on 3D CT images by using deep learning, a completely novel approach, instead of conventional segmentation methods originated from traditional digital image processing techniques. We evaluated and compared the segmentation performances of two different deep learning approaches based on 2D- and 3D deep convolutional neural networks (CNNs) without and with a pre-processing step. A conventional method based on a probabilistic atlas algorithm, which presented the best performance within the conventional approaches, was also adopted as a baseline for performance comparison. A dataset containing 240 CT scans of different portions of human bodies was used for training the CNNs and validating the segmentation performance of the learning results. A maximum number of 17 types of organ regions in each CT scan were segmented automatically and validated with the human annotations by using ratio of intersection over union (IoU) as the criterion. Our experimental results showed that the IoUs of the segmentation results had a mean value of 79% and 67% by averaging 17 types of organs that were segmented by the proposed 3D and 2D deep CNNs, respectively. All results using the deep learning approaches showed better accuracy and robustness than the conventional segmentation method that used the probabilistic atlas algorithm. The effectiveness and usefulness of deep learning approaches were demonstrated for multiple organ segmentation on 3D CT images.

Keywords

CT image · Multiple organs · Image segmentation · Deep learning · CNN · Patch-based training · 2D-FCN · Majority voting

X. Zhou (✉)
Gifu University, Gifu-shi, Gifu, Japan
e-mail: zxr@gifu-u.ac.jp

G. Lee, H. Fujita (eds.), *Deep Learning in Medical Image Analysis*, Advances in Experimental Medicine and Biology 1213, https://doi.org/10.1007/978-3-030-33128-3_9

Introduction

CT image segmentation, annotation of the target organ regions on CT images, is a fundamental step of the medical image analysis that is required for diagnosis, surgery, and therapy [1]. It is also one of the most labor-intensive steps for clinicians. Segmentation of all organ regions within a 3D CT image is a critical processing step for realizing precision and personalized medicines. Although the research of CT image segmentation had a long history [2–8], the conventional approaches originated from digital image processing techniques still showed weak performances on accuracy, efficiency, and robustness, especially on automatically segmenting multiple organs on the CT images scanned for a wide range of human body. Recently, deep learning approaches demonstrated a high performance of scene segmentation on pictures taken by video cameras [9, 10], and showed the potentials to address organ segmentation problem on medical images [11–15]. However, there are still very few comprehensive evaluations of segmentation performance of deep-learning-based approaches on simultaneously segmenting all of the major organs on 3D CT images, which may be scanned on various portions of human body.

CT image segmentation involves a wide range of research topics that cannot be all covered in this chapter. Therefore, we limit our discussion on one of the most challenging issues, simultaneous segmentations of multiple organs by using deep learning [16], which can be simply defined as a voxel-based annotation for all major organ types on a 3D CT image that may present different ranges of the human body. Generally, the performance of a deep-learning-based approach highly depends on three factors: (1) training dataset, (2) model size, and (3) computational capability (performance and resource of graphics processing unit: GPU), thus our discussion about deep-learning-based CT image segmentation starts from these factors. In the following, we first briefly introduce the current status on these factors from the perspective of demand on CT image segmentation, and then we describe two deep learning approaches [16–20] based on 2D and 3D deep convolutional neural networks (CNNs) without and with a pre-processing step to accomplish multiple organ segmentation tasks. Finally, we show a validation result by using deep learning approaches on a CT image dataset, which contains 240 CT scans covered different portions of human bodies with the accurate pixel-wise annotations of 17 types of major organs from radiologists. The results by using a conventional method [7] that presented the best performance of the previous generation of methodology besides deep learning are also presented as a baseline for performance comparisons.

Issue of Deep Learning for CT Image Segmentation

The input of a segmentation process is a 3D CT image, and the output is a label map of the same size and dimension, in which the anatomical structures are annotated by a pre-defined set of labels based on human anatomy. The segmentation process is simply repeating annotation for each voxel of the CT images by using a pre-defined rule that is either defined manually by human experience or learned automatically from known CT scans by a computer. Deep learning acts as a rule-generator and stores the rule as the parameters of a CNN to accomplish the voxel-wise annotation. The parameters of the CNN are learned by trial-and-error on a training dataset with the known annotations. The structure (called hyper-parameter) of the CNN is decided by human design in most cases and never changed during the training stage. In general, a bigger model (deeper CNN structure with more parameters) trained on a bigger dataset (a larger number of CT scans with accurate annotations) potentially has a better performance on image segmentation. However, the growth of the size of the model and dataset is constrained by the compute capability and resource of the computational hardware.

A 3D CT scan is a volumetric image data and constructed by a series of 2D CT slices. Each

CT slice is a 512 × 512 image matrix where the CT numbers are stored in a format of two-byte integer. A typical CT scan for human torso includes about 1024 slices and consumes memory of 512 Mbytes in a computer or GPU. This data size is 5000 times larger than the assumption in most deep learning tasks (a typical input picture is 24 bits of RGB colors stored in a 224 × 224 matrix and consumes 147 Kbytes of memory). On the other hand, the models used in most research works for CT image segmentation are based on the CNN architectures that originally proposed for 2D picture recognition in computer version tasks with the same or larger model size. For example, two models (3D-Unets [12] and V-net [13]) specialized to accomplish 3D medical image segmentations have about 100M learning parameters, which is comparable to the other CNN models used in computer vision research field.

Unfortunately, the current capability of computational hardware (memory size of GPU) is not sufficient enough to store and process the whole 3D CT image directly for training a deep CNN model to accomplish a wide range of anatomical structures recognition. A compromise way is to divide a 3D CT image into a number of sub-regions (2D or 3D image patches in a small data size) and train a CNN model to recognize a large number of fragments of human anatomy appeared in those sub-regions. However, dividing a CT image into a number of small pieces leads to a loss of the global information of anatomical structures and increases the variety of image appearances in the training dataset that affects the performance of deep learning. Therefore, how to decompose a CT scan into a number of small image patches to enable deep learning for CT image segmentation based on current GPUs is the first issue that needs to be clarified before considering the techniques of deep learning.

We proposed two approaches to address the issue mentioned above, which decompose the 3D CT scans into patches and apply modern deep CNNs to accomplish the multiple organ segmentation task. In the next section, we will briefly introduce our previous works [16, 17] and show the validation results of the segmentation performances based on a CT image dataset.

Two Approaches for Multiple Organ Segmentations Using 2D and 3D Deep CNNs on CT Images

Overview

Two approaches (a) decomposing a 3D CT scan into a number of 2D sections from different orientations [16] and (b) detecting the 3D-bounding box [21] of each organ and scaling these image regions within the bounding boxes into a fix-sized 3D-region-of-interest (ROI) are proposed to enable deep learning approach [17] to accomplish CT image segmentation based on current GPU hardware. Two modern CNN structures, fully convolutional network (FCN) [9], and Vnet [13] that is based on residual network (ResNet) [10] are used as the templates to construct our CNN networks. Two loss functions, pixel-based labeling loss and region-based coincidence (Dice index) between the segmentation result and the human annotation, are used to guide the training process of the CNNs. Two optimization methods, stochastic gradient descent (SGD) and ADAM [22], were used for learning the network parameters. Details are described in the following sections.

Deep Learning Anatomical Structures on 2D Sectional Images

As the core part of the approach (a), a 2D deep learning method was applied to multiple organ segmentation on each 2D section of 3D CT scans [18–20]. Our approach modeled the CT image segmentation as "multiple 2D proposals with 3D integration" and got inspiration from the behavior that radiologists interpret a CT scan from many 2D sections on a screen and reconstruct 3D anatomical structures mentally [20]. A 2D fully convolutional network (FCN) [9] extended from a VGG-16 network was used for generating the multiple 2D proposals of organ regions from three body directions on each CT scan independently, and an organ–label fusion (majority vote of the proposals) was carried out on each voxel in 3D

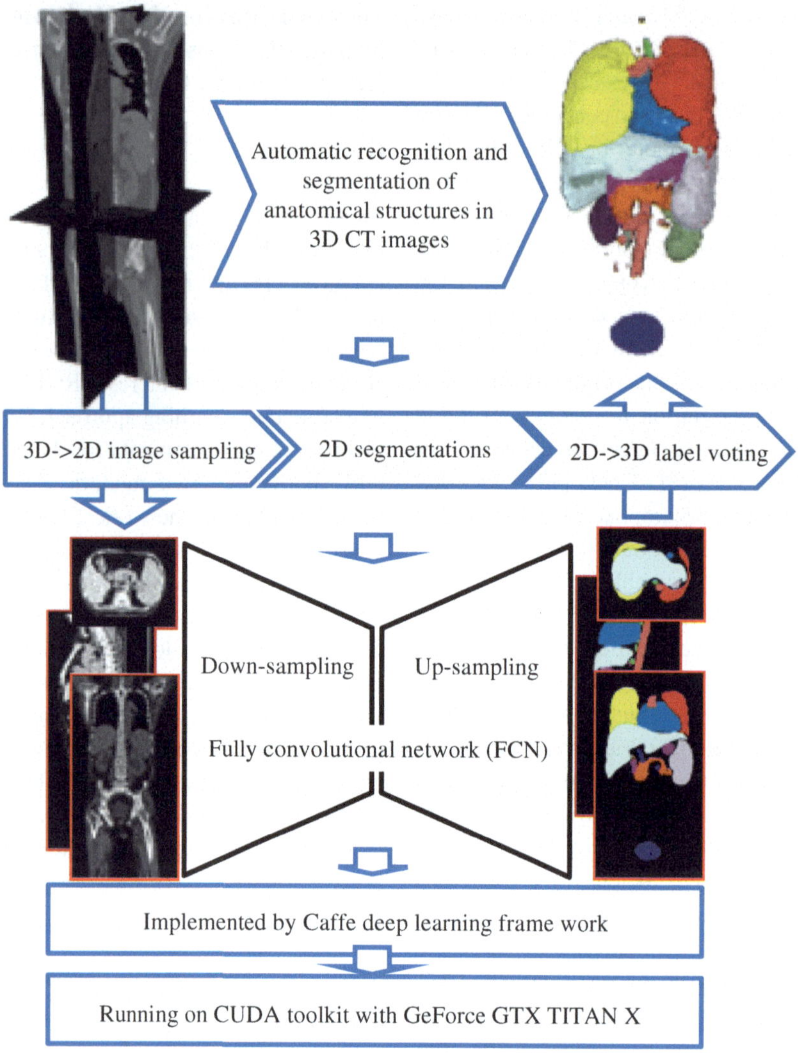

Fig. 1 The process flow of multiple organ segmentations on 3D CT images by using 2D FCN with a majority voting [16]

CT image space [18]. In the training stage, we decompose the 3D CT scans with annotated organ regions into several 2D sections in three orthogonal body directions and treat each 2D slice with an annotation as a training sample. The sum of the pixel-wise classification error on 2D slices was used as the loss function during the training process. ADAM [22] was used as the optimization method. In the testing stage, we applied the trained FCN to segment partial organ regions on each 2D section along three body directions independently, and then integrated the segmented 2D partial organ regions into 3D CT image space by using a voxel-wise label fusion to obtain the final 3D segmentation result as shown in Fig. 1. The implementation details of the processing steps and network structures can be found in [16].

Deep Learning Local Appearances of Multiple Organs on 3D CT Images

We proposed a 3D deep learning approach for multiple organ segmentation [17]. Our approach

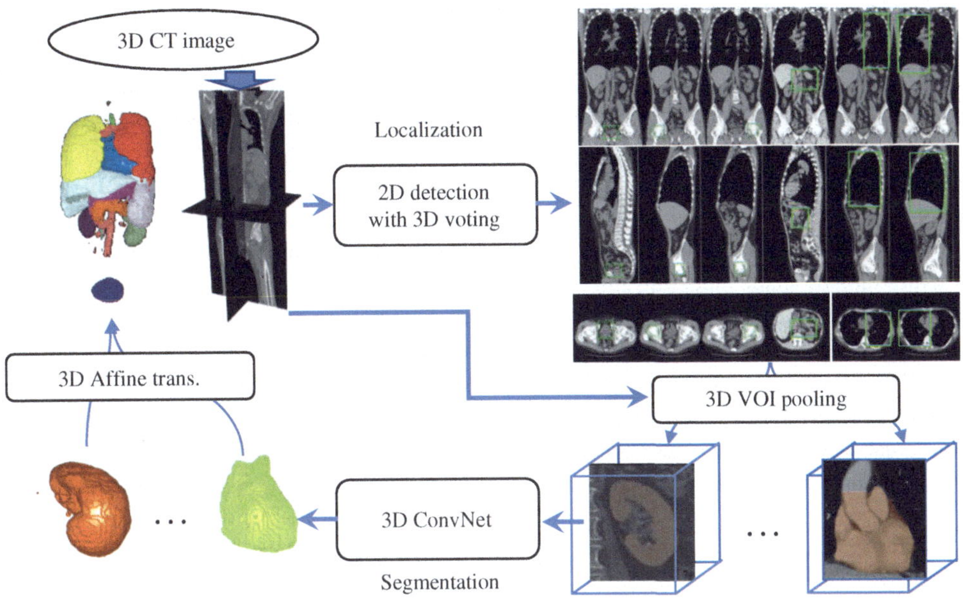

Fig. 2 The process flow of multiple organ segmentations by combining organ localization and organ segmentation from CT image [17]

accomplished organ segmentation through two steps, as shown in Fig. 2. We decoupled the organ detection and segmentation functions, and modeled the multiple organ segmentation processes as "detecting the bounding box of different organs first, and then segmenting the organ region within each 3D bounding box." The organ detection module used in this work was presented in our previous work [21]. This method used a conventional machine-learning approach based on window sliding and pattern matching in Haar-like and LBP image feature spaces through the entire CT image space to detect each organ region and return the coordinates of a bounding box as the organ location. Details of our organ detection method can be found in [21]. The organ segmentation module used a 3D deep CNN structure derived from VNet [12, 13]. This 3D deep CNN was constructed using a number of 3D convolutional layers based on ResNet [10] modules with skip-connections to combine the information in shallow and deep layers. The Dice index between the segmented region and ground truth was used as the loss function, and the SGD algorithm was used for optimizing the network parameters. In the training stage, we trained organ localization and segmentation separately. We generated the ground truth of the bounding box for each organ type from the manual annotations in each CT scan and trained a set of organ-specific ad-hoc classifiers that detect the bounding box for each organ type in CT scans. Next, we cropped the volume-of-interest (VOI) regions for each organ type based on the bounding box and resized the VOIs to the same image size ($128 \times 128 \times 128$ voxels) as the input image of the segmentation model [17]. The annotation of each organ region was also cropped and resized in the same manner to retain the correspondence of the spatial location of each voxel between the CT image and manual annotation data. The pairs of VOI and annotation of all organ regions in CT scans were used as the samples for training the 3D deep CNN. In the testing stage, our method detected the bounding boxes of all the organ regions within a CT scan and segmented target organ regions within the range of bounding boxes [17].

Conventional Image Segmentation Approach

A conventional method presented by Okada et al. [7] was used as the baseline for evaluating the performance of the deep learning approaches. This conventional method modeled the intensity (CT number) and shape information of each target organ by using statistic shape modeling and probabilistic atlas, and combined intra-organ information with an inter-organ relationship to guide the segmentation process. This method was relatively new and demonstrated the best performance for abdominal organs segmentation than the other conventional methods based on a large CT image dataset [7].

Results

The performance evaluations were based on a shared dataset produced by a research project titled "Computational Anatomy [23]." This dataset consists of 240 CT scans from 200 patients for the diagnosis of diverse types of lesions, obtained at Tokushima University Hospital. These CT images were scanned for different portions (89 torso, 17 chest, 114 abdomen, and 20 abdomen-with-pelvis scans) by a multi-slice CT scanner (Aquilion from Toshiba Medical Systems Corporation) and reconstructed by different protocols and kernel functions, leading to different image resolutions (distributed between 0.625 and 1.148 mm with 1.0 mm slice thickness) and different image quality (specialized for lung or abdominal observations). Contrast media enhancements were applied in 155 CT scans. The anatomical ground truth (a maximum of 19 labels that show 17 major organ types and 2 interesting regions inside the human body) of 240 CT scans was manually annotated [24] and distributed within the dataset. Here, we only consider the 17 organ regions as the segmentation targets for performance evaluations [17].

We directly used the organ localization module that was generated based on another CT image dataset in our previous work [21] without any fine-tuning to adapt to above CT image database. Because the segmentation approaches used in this work were based on machine learning, we used a fourfold cross-validation for performance evaluation. We used 75% CT scans of the dataset for training parameters of the different network structures and tested the trained networks on the remaining 25% CT scans. CT scans from the same patient were only used in the same training or testing stage. The accuracy of the segmentation was evaluated for each organ type and each CT scan. The intersection over union (IoU) (also known as the Jaccard similarity coefficient) between the segmentation result and ground truth was used as the evaluation measures. Because a CT scan may contain different organ regions, we used a comprehensive evaluation of multiple organ segmentation results for all CT scans by considering the variance of the organ number and volume. The measures (mean voxel accuracy, mean IoU, and frequency-weighted IoU) that are commonly used in semantic segmentation and scene parsing [9] were employed in this study for the evaluations. Let n_{ij} be the number of pixels in target i classified as target j, n_{cl} be the total number of different targets in a CT case, and $t_i = \sum_j n_{ij}$ be the total number of pixels in target i. These measures are defined as:

- Mean voxel accuracy:

$$\left(\sum_i n_{ii}/t_i\right)/n_{cl} \quad (1)$$

- Mean IoU:

$$\left(\sum_i n_{ii}/\left(t_i + \sum_j n_{ji} - n_{ii}\right)\right)/n_{cl} \quad (2)$$

- Frequency-weighted IoU:

$$\left(\sum_k t_k\right)^{-1} \sum_i t_i n_{ii}/\left(t_i + \sum_j n_{ji} - n_{ii}\right) \quad (3)$$

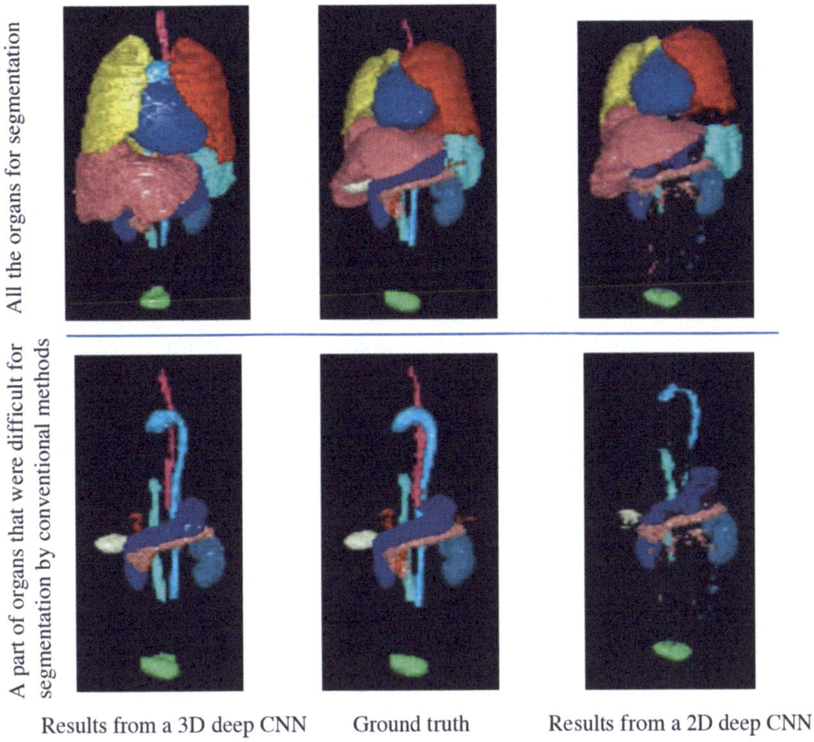

Fig. 3 3D views of the anatomical structure segmentation results in a CT scan by using 3D (left side) and 2D (right side) deep CNN approaches with the human annotation (middle column) as the ground truth [17]

An example of the result of multiple organ segmentation in a testing CT scan is shown in Fig. 3 by using a 3D volume rendering method. The evaluation results of deep CNNs demonstrated that the average segmentation accuracy of 17 types of organ over the 240 (4 × 60) test CT scans was 67% (2D CNN) and 78.8% (3D CNN) in terms of the mean IoUs (refer to Table 1), 84.9% (2D CNN) and 89% (3D CNN) in terms of the frequency-weighted IoUs, and 86.1% (2D CNN) and 88.4% (3D CNN) in terms of the mean voxel accuracy [17].

We tested the conventional method presented by Okada et al. [7] based on the same training and testing CT images by using their system for both model construction and organ segmentation. Because this conventional method was only applied to abdominal CT scans, our experiment was limited to seven organ types including liver, gallbladder, left and right kidneys, spleen, pancreas, and veins. The experimental results showed that the average segmentation accuracy of seven types of organs was 59% in terms of the mean IoUs, which was lower than the results of the 2D and 3D CNN approaches [17].

The deep CNNs used in this work were implemented based on Caffe deep learning framework [25]. The computing time for training a 3D deep CNN was about three days. This was longer than the time taken for training a 2D deep CNN network, which is approximately one and a half days based on a GPU (NVIDIA Quadro P6000 with 24 GB of memory). The organ segmentation for one testing 3D CT scan with a 512 × 512 × 221 matrix took approximately 3 min by using a trained deep CNN. The efficiency in terms of system development and improvement of 2D and 3D deep CNN approaches was better than that of the conventional method that used a CPU based approach [17].

Table 1 Accuracy evaluations in terms of mean value of IoUs for 17 target types between segmentation and ground truth in 60 test CT cases [17]

Organ type	3D-Deep CNN	2D-Deep CNN
Right lung	95.1	93.9
Left lung	94.4	93.5
Heart	89.0	86.0
Aorta	81.0	72.5
Esophagus	57.4	34.7
Liver	91.1	90.8
Gallbladder	80.6	61.3
Stomach	56.8	46.0
Spleen	91.3	83.5
Right kidney	89.6	83.7
Left kidney	89.6	82.4
Inferior vena cava	74.3	55.6
Portal vein	58.7	33.3
Pancreas	65.7	46.0
Bladder	86.7	68.1
Prostate	74.2	47.0
Uterus	63.7	42.7

Discussions

Segmentation Performances

We confirmed that the most anatomical structures were segmented correctly by using deep CNN based approach except for few very small organs such as gallbladder and prostate. 3D deep CNN highly related on the bounding box detected by the pre-processing and showed relatively better IoUs than 2D deep CNN method. However, we found the 2D deep CNN was simpler and more practical. It automatically adapted to the different portions of the CT scans and successfully segmented target organs within the CT images without any need of operator intervention or predefined parameters that was impossible in the previous works. Three examples of using the 2D deep CNN to accomplish the multiple organ segmentations on torso, chest, abdomen CT scans were shown in Fig. 4. Because the segmentation targets cover a wide range of shapes, volumes, and sizes, either with or without contrast enhancement, and come from different locations in the human body, these experimental results offer an excellent demonstration of the capability of deep CNN approaches to recognize the major anatomical structures in the types of CT images actually used in clinical medicine [16].

We investigated the mean IoUs of each organ type by using 2D and 3D deep CNN. The IoUs of the segmentation results have a relation to the volume of the target region. We found the IoUs of the organs with larger volumes (e.g., lung, liver) were satisfied (IoUs were greater than 90%) and differences between the segmentation results by 2D and 3D deep CNN were not large. However, for some smaller organs (e.g., gallbladder, uterus) and stomach contents (which have not previously been reported), the 3D deep CNN produced a relatively higher IoU than 2D deep CNN, but the performance was still not good enough (IoUs were less than 70%). The significant degradation on IoU (20% or more) may occur on three reasons: (1) The spatial resolution of CT images was not enough to provide rich image textures for the deep learning to find the discriminative features for recognition and segmentation of such small regions. (2) The lower pixel-occupancy of such small organ regions on CT images caused a smaller weight in loss value during the training process, which decides the importance and priority of target organ during the segmentation process. That's means the small organs sometimes were ignored by the segmentation process. (3) The instruction signals for small organs with a 3D tube- or line-shape were relatively expensive to be annotated manually and had a smaller number in our dataset, which may cause an insufficient learning for segmenting such kinds of organs. In our approach based on 3D deep CNN, we used a pre-processing (detecting bounding box to balance the volumes of background and target organ on input images and normalize the spatial resolution of each organ type, using a data augmentation to balance the numbers of different organ types in training dataset) to address these concerns described above, and confirmed a significant improvement on IoUs on such small organ types comparing to the results from the 2D deep CNN [17].

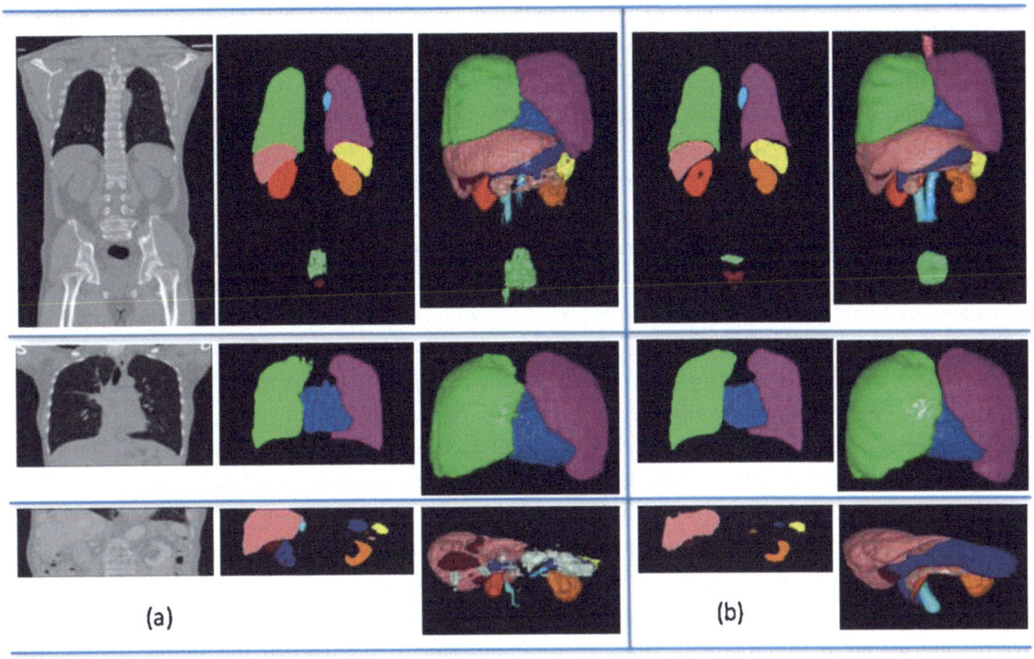

Fig. 4 (**a**) Three examples of segmentation by using a 2D deep CNN in 3D CT case covering torso (upper), chest (middle), and abdomen (lower) regions along with segmented regions labeled with different colors for one 2D coronal CT slice (middle column) and 3D visualization based on surface-rendering method (right column). (**b**) Corresponding ground-truth segmentations for three cases [16]

Training Protocol and Transfer Learning

In our paper [16], we compared two different training protocols for 2D deep CNN for CT image segmentation. We firstly trained a 2D CNN network based on "learn from scratch" that was simply initializing all parameters of 2D deep CNN to a small number with zero mean and confirmed no convergence was observed within 80,000 learning iterations, and the network then failed to segment any of the targets in the test dataset. And then, we fine-tuned our network using which is pretrained using ImageNet [26], convergence was achieved after 22,000 iterations. The trained network fine-tuned in 80,000 learning iterations could segment multiple organs successfully in CT images from both the testing and training datasets. This demonstrates that comprehensive knowledge learned from large-scale, well-annotated datasets [26] can be transferred to our network to accomplish the CT image segmentation task.

In the case of training 3D deep CNN, we have to "learn from scratch" because there was no pre-trained model that can be used as the start point for fine-tuning the 3D CNN networks [17]. We tried to use modern initialization methods [27, 28] at the beginning of training our 3D deep CNN. The experimental result demonstrated that the convergence of our 3D deep CNN was observed after 5000 learning iterations and the segmentation performance was stable and satisfied at 40,000 learning iterations. A suitable initialization was the key to accomplish a successful deep learning for CT image segmentation. We found that these novel initialization methods [27–29] were also helpful for training a 2D deep CNN from scratch. The transfer learning based on pre-trained models was helpful for training process to reach a convergence of the loss values quickly and output a useful result, but it did not affect the final performance of the trained CNN to accomplish multiple organ segmentation on CT images in our works.

We compared the performance of 2D deep CNN networks optimized by SGD and ADAM methods with the same training protocols. The segmentation results on the test data indicate that the network trained by ADAM offers slightly better performance (up by 0.3% in voxel accuracy, 0.15% in frequency-weighted IoU) than that trained by SGD. Because the learning rate does not need to be tuned in ADAM and the default parameters are likely to achieve good results [16].

The performance of the CNN network was affected by the number of training iterations. We compared the segmentation results on the test dataset given by 2D deep CNN networks after 80,000, 160,000, and 550,000 training iterations. We found that 160,000 iterations were sufficient to train the network. Further training iterations may improve the segmentation accuracy of some organ types, but could not improve the overall performance across all organ types [16].

Comparison to Conventional Methods

The previous studies most closely related to our work are those of Udupa et al. [8], Wolz et al. [6], Okada et al. [7], and Lay et al. [5]. Common to all these works is that they focused on multiple organ segmentation in CT images, as in our study [16, 17]. Among these works, the coarse-to-fine approach has been used and demonstrated a good performance. A typical scheme of this coarse-to-fine approach is to recognize the rough location of a target organ firstly and then extracts the accurate organ contour to separate the target region from the background on CT images. The core part is to generate some models (e.g., fuzzy model, probabilistic atlas, active shape model) manually to store distributions of handcrafted features, which relate to anatomy within a training dataset, and use these models to guide/constrain the conventional image processing to accomplish multiple organ segmentations.

We broke down the black-box of learned 2D deep CNN and aimed to interpret what was inside the training result (the contents that learned from the CT dataset) [16]. We surprisingly found that the learned deep CNN also naturally followed a coarse-to-fine approach to accomplish the image segmentation as same as the previous works described above [16]. The 2D deep CNN structure used for image segmentation in this work was constructed by two connected blocks: down-sampling path and up-sampling path (Fig. 5) with a skip-connection [9, 12]. We found that the CNN layers in down-sampling path accomplished the recognition of different organs and outputted a rough location (score of down-sampling path at lower part in Fig. 5) of each target region on a CT image, which likes the function of probabilistic atlas used in conventional methods. The layers in up-sampling path of CNN gradually recovered the contour of each target organ (refer to upper part in Fig. 5), which likes the conventional contour decisions. Because the contour presents the shape of the target organ, we can say that the deep CNN learned the organ shapes from the training CT images and use it to recover the contour of the target organ regions on an unseen CT image, which was just the aim and usage of the shape modeling used in the previous works. This result demonstrated that the deep CNN implicitly learned the same policy that had been proved useful for CT image segmentation in conventional methods based on the human experience.

Our experimental results showed that the CNN models had much better flexibility and capability to fit the variance of organ appearances than the hand-crafted modes [7], which caused a few failures (IoU was equal to zero) for some organs segmentation in our testing stage. The improvements on generality of CT image segmentation by using deep CNNs were demonstrated on the test CT scans. We believe this improvement was because of the model size (parameters) of the deep CNN was much bigger than the handcrafted models in conventional method [7] and more CT scans could be used for training the model than previous works [7]. We found that training a CNN model took more computing time (daily basis on GPU) than manually generating the atlas and shape models (hourly basis on CPU) in previous work [7], large model, efficient model optimization with more CT scans for model training

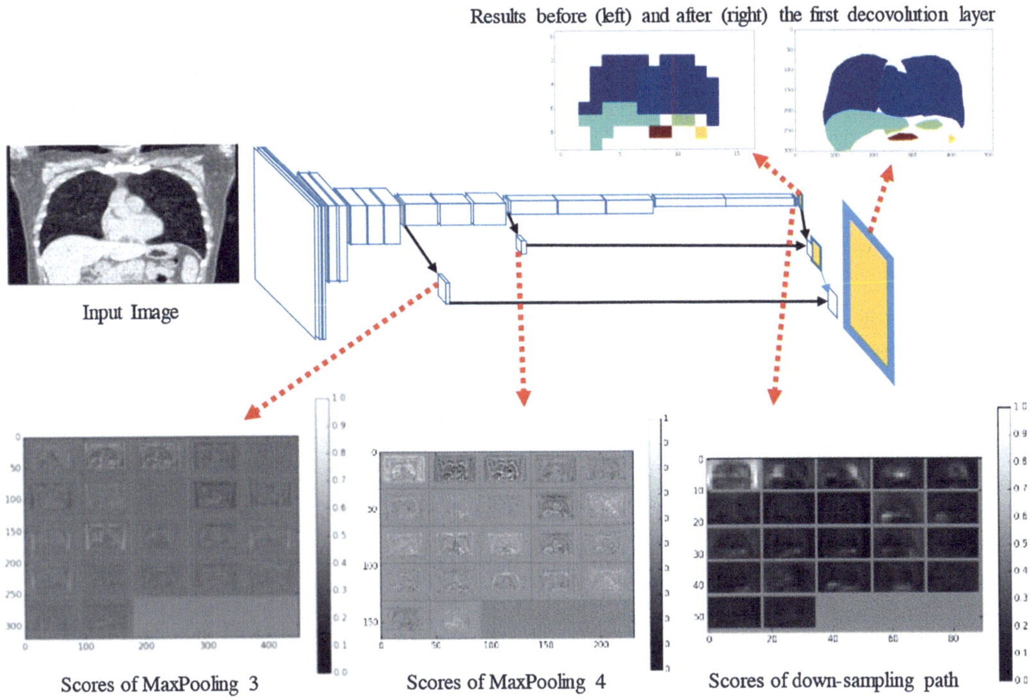

Fig. 5 Insight of learned 2D deep CNN [16]. Lower: outputs of middle layers and final result of down-sampling path; Upper: outputs before and after the first deconvolution layer

demonstrated the advantage of deep learning for CT image segmentation.

We compared the IoUs of champion data on major organ types segmented by 2D deep CNN and the conventional method [7] that used handcrafted models. We could not observe a significant difference. In the cases that only a very small dataset can be used for system design, the computation capability of the hardware is not strong enough, and the target organ for segmentation has a relatively small variance on CT image appearance, the conventional approach based on handcrafted models may be still useful and practical for developing organ segmentation scheme on CT images.

Computational Efficiency

Deep CNN based segmentation process showed a high computational efficiency because of its simple structure (matrix calculations) that efficiently operated by GPU-based implementation. The computing time for training a 3D deep CNN was about three days. This was longer than the time taken for training a 2D deep CNN network, which is approximately one and a half days based on a GPU (NVIDIA Quadro P6000 with 24 GB of memory). The organ segmentation for one testing 3D CT scan with a 512 × 512 × 221 matrix took approximately 3 min by using a trained 3D deep CNN [17]. In the case of 2D deep CNN, the computing time for multi-organ segmentation of one 2D CT slice takes approximately 30 ms. These computing times were much faster than the conventional method [7] to segment one abdominal CT scan based on a CPU (Intel core i7).

We found that the efficiency of deep CNN based approach in terms of system development and improvement was much better than that of previous works that attempted to incorporate human specialist experience into complex algorithms for segmenting different organs [16]. Furthermore, neither the target organ type, number of organs within the image, nor image size limits the CT images that are used for the training process. Although labeling the anatomical structures in CT images as the training samples still takes time, this burden can be reduced using bespoke and advanced semi-automatic algorithms [16].

Conclusion

We proposed two approaches [16, 17] to apply modern deep learning techniques to recognize and segment anatomical structures from 3D CT images based on current computer hardware (GPU). In our methods, 3D CT images are decomposed into two kinds of image patches with small image size and they are fed to 2D and 3D deep CNNs to accomplish the segmentations for 17 types of organ regions on 240 CT scans. The accuracy of the segmentation results by using these approaches was evaluated based on fourfold cross-validation and compared with a conventional method.

We confirmed that (1) our proposed approaches combined with deep learning demonstrated a better performance (especially in terms of robustness and generality) for CT image segmentations than the conventional method. (2) Deep CNN showed capability to learn the fractional appearance of the 3D anatomical structures on CT image patches that had a large variance based on relatively a small number of CT scans. (3) The performances of 2D and 3D CNN were comparable for segmenting the massive organ types with a large volume. A 3D CNN showed a better accuracy of segmentation for small volume organ types with a tube- or line-shape [17].

In conclusion, the deep learning approaches showed better accuracy and efficiency for multiple organ segmentations on 3D CT images than the conventional approaches. The issue of CT image segmentation may be addressed more successfully by using deep learning with the growth of the computational capability and more CT data for training in the future.

Acknowledgments The author would like to thank all the members of the Fujita Laboratory in Gifu University, for their collaboration. We especially thank Dr. Okada of Tsukuba University for providing binary code and Dr. Roth of Nagoya University for giving advice on 3D deep CNN. We would like to thank all the members of the "Computational Anatomy" research project [22], especially Dr. Ueno of Tokushima University, for providing the CT image database. This research was supported in part by a Grant-in-Aid for Scientific Research on Innovative Areas (Grant No. 26108005), and in part by a Grant-in-Aid for Scientific Research (C26330134), MEXT, Japan.

References

1. Doi K (2007) Computer-aided diagnosis in medical imaging: historical review, current status, and future potential. Comput Med Imaging Graph 31:198–211
2. Pham DL, Xu C, Prince JL (2000) Current methods in medical image segmentation. Biomed Eng 2:315–333
3. Heimann T, Meinzer HP (2009) Statistical shape models for 3D medical image segmentation: a review. Med Image Anal 13(4):543–563
4. Xu Y, Xu C, Kuang X, Wang H, Chang EIC, Huang W, Fan Y (2016) 3D-SIFT-Flow for atlas-based CT liver image segmentation. Med Phys 43(5):2229–2241
5. Lay N, Birkbeck N, Zhang J, Zhou SK (2013) Rapid multi-organ segmentation using context integration and discriminative models. Proc IPMI 7917:450–462
6. Wolz R, Chu C, Misawa K, Fujiwara M, Mori K, Rueckert D (2013) Automated abdominal multi-organ segmentation with subject-specific atlas generation. IEEE Trans Med Imaging 32(9):1723–1730
7. Okada T, Linguraru MG, Hori M, Summers RM, Tomiyama N, Sato Y (2015) Abdominal multi-organ segmentation from CT images using conditional shape-location and unsupervised intensity priors. Med Image Anal 26(1):1–18
8. Sun K, Udupa JK, Odhner D, Tong Y, Zhao L, Torigian DA (2016) Automatic thoracic anatomy segmentation on CT images using hierarchical fuzzy models and registration. Med Phys 43(4):1882–1896
9. Long J, Shelhamer E, Darrell T (2015) Fully convolutional networks for semantic segmentation. Proc CVPR:3431–3440
10. He K, Zhang X, Ren S, Sun J (2016) Deep residual learning for image recognition. Proc CVPR:770–778
11. Shin HC, Roth HR, Gao M, Lu L, Xu Z, Nogues I, Yao J, Mollura D, Summers RM (2016) Deep convolutional neural networks for computer-aided detection: CNN architectures, dataset characteristics and transfer learning. IEEE Trans Med Imaging 35(5):1285–1298
12. Ronneberger O, Fischer P, Brox T (2015) U-Net: convolutional networks for biomedical image segmentation. In: Proceedings of the Medical Image Computing and Computer-Assisted Intervention (MICCAI), pp 234–241

13. Milletari F, Navab N, Ahmadi S-A (2016) V-Net: fully convolutional neural networks for volumetric medical image segmentation. arXiv:1606.04797
14. Roth HR, Farag A, Lu L, Turkbey EB, Summers RM (2015) Deep convolutional networks for pancreas segmentation in CT imaging. Proc SPIE 9413:94131G-1–94131G-8
15. Cha KH, Hadjiiski L, Samala RK, Chan HP, Caoili EM, Cohan RH (2016) Urinary bladder segmentation in CT urography using deep-learning convolutional neural network and level sets. Med Phys 43(4):1882–1896
16. Zhou X, Takayama R, Wang S, Hara T, Fujita H (2017) Deep learning of the sectional appearances of 3D CT images for anatomical structure segmentation based on an FCN voting method. Med Phys 44(10):5221–5233
17. Zhou X, Yamada K, Kojima T, Takayama R, Wang S, Zhou XX, Hara T, Fujita H (2018) Performance evaluation of 2D and 3D deep learning approaches for automatic segmentation of multiple organs on CT images. In: Proceedings of the SPIE Medical Imaging 2018: Computer-aided diagnosis, 105752C
18. Zhou X, Ito T, Takayama R, Wang S, Hara T, Fujita H (2016) Three-dimensional CT image segmentation by combining 2D fully convolutional network with 3D majority voting. In: Proceedings of the Workshop on the 2nd Deep Learning in Medical Image Analysis (DLMIA) in MICCAI 2016, LNCS 10008, pp 111–120
19. Zhou X, Takayama R, Wang S, Zhou X, Hara T, Fujita H (2017) Automated segmentation of 3D anatomical structures on CT images by using a deep convolutional network based on end-to-end learning approach. In: Proceedings of the SPIE Medical Imaging 2017: Image Processing, vol 10133, p 1013324
20. Zhou X, Ito T, Takayama R, Wang S, Hara T, Fujita H (2016) First trial and evaluation of anatomical structure segmentations in 3D CT images based only on deep learning. Brief Article. Med Image Inf Sci 33(3):69–74
21. Zhou X, Morita S, Zhou XX, Chen H, Hara T, Yokoyama R, Kanematsu M, Hoshi H, Fujita H (2015) Automatic anatomy partitioning of the torso region on CT images by using multiple organ localizations with a group-wise calibration technique. In: Hadjiiski LM, Tourassi GD (eds) Proceedings of the SPIE Medical Imaging 2015: Computer-Aided Diagnosis, vol 9414, pp 94143K-1–94143K-6
22. Kingma DP, Ba JL (2015) ADAM: a method for stochastic optimization. In: Proceedings of the ICLR 2015
23. http://www.comp-anatomy.org/wiki/
24. Watanabe H, Shimizu A, Ueno J, Umetsu S, Nawano S, Kobatake H (2013) Semi-automated organ segmentation using 3-dimensional medical imagery through sparse representation. Trans Jpn Soc Med Biol Eng 51(5):300–312
25. http://caffe.berkeleyvision.org
26. http://www.image-net.org
27. Glorot X, Bengio Y (2010) Understanding the difficulty of training deep feedforward neural networks. In: International Conference on Artificial Intelligence and Statistics
28. He K, Zhang X, Ren S, Sun J (2015) Delving deep into rectifiers: surpassing human-level performance on ImageNet classification. arXiv preprint arXiv:1502.01852
29. Ioffe S, Szegedy C, Batch normalization: accelerating deep network training by reducing internal covariate shift. arXiv preprint arXiv:1502.03167

Techniques and Applications in Skin OCT Analysis

Ai Ping Yow, Ruchir Srivastava, Jun Cheng, Annan Li, Jiang Liu, Leopold Schmetterer, Hong Liang Tey, and Damon W. K. Wong

Abstract

The skin is the largest organ of our body. Skin disease abnormalities which occur within the skin layers are difficult to examine visually and often require biopsies to make a confirmation on a suspected condition. Such invasive methods are not well-accepted by children and women due to the possibility of scarring. Optical coherence tomography (OCT) is a non-invasive technique enabling in vivo examination of sub-surface skin tissue without the need for excision of tissue. However, one of the challenges in OCT imaging is the interpretation and analysis of OCT images. In this review, we discuss the various methodologies in skin layer segmentation and how it could potentially improve the management of skin diseases. We also present a review of works which use advanced machine learning techniques to achieve layers segmentation and detection of skin diseases. Lastly, current challenges in analysis and applications are also discussed.

A. P. Yow · D. W. K. Wong (✉)
Institute for Health Technologies, Nanyang Technological University, Singapore, Singapore

SERI-NTU Advanced Ocular Engineering (STANCE), Singapore, Singapore

Singapore Eye Research Institute, Singapore National Eye Centre, Singapore, Singapore

R. Srivastava
Institute for Infocomm Research, A * STAR, Singapore, Singapore

J. Cheng
Cixi Institute of Biomedical Engineering, Chinese Academy of Sciences, Beijing, China

A. Li
Beihang University, Beijing, China

J. Liu
Cixi Institute of Biomedical Engineering, Chinese Academy of Sciences, Beijing, China

Southern University of Science and Technology, Shenzhen, China

L. Schmetterer
SERI-NTU Advanced Ocular Engineering (STANCE), Singapore, Singapore

Center for Medical Physics and Biomedical Engineering, Medical University of Vienna, Vienna, Austria

Department of Clinical Pharmacology, Medical University of Vienna, Vienna, Austria

H. L. Tey
National Skin Centre, Singapore, Singapore

Yong Loo Lin School of Medicine, National University of Singapore, Singapore, Singapore

Lee Kong Chian School of Medicine, Nanyang Technological University, Singapore, Singapore

G. Lee, H. Fujita (eds.), *Deep Learning in Medical Image Analysis*, Advances in Experimental Medicine and Biology 1213,
https://doi.org/10.1007/978-3-030-33128-3_10

Keywords
Skin · Optical coherence tomography (OCT) · Roughness · Dermal-epidermal junction (DEJ) · Basal cell carcinoma (BCC) · Epidermis · Dermis · Segmentation · Graph · Deep learning

Introduction

Skin is the largest organ of the integumentary system which acts as an outer covering of our body. It is composed of three main layers, namely epidermis, dermis and hypodermis. Epidermis is the outermost layer that protects the body from the environment. Its varying thickness depends on factors such as anatomical sites, age and skin diseases. The dermis is the next layer located beneath the epidermis and contains number of skin cells and structures such as sweat glands, hair follicles and blood vessels. The last layer, known as the hypodermis, lies below the dermis and is used mainly for fat storage and functions as insulation and padding for our body (Fig. 1).

Skin disease abnormalities mostly occur within these layers, which is impossible to examine with naked eye. Hence, histological analysis through a biopsy is often required to identify presence of abnormal lesions. Due to the invasive nature of such analysis, both clinicians and researchers have keen interest in the use of non-invasive approaches for skin diagnosis. Optical coherence tomography (OCT) is one of the popular non-invasive imaging methods that enables examination of sub-surface skin tissue without the need for biopsy. The OCT imaging technique was first used in the field of ophthalmology, and proved to be of value for visualizing changes in retina and cornea [1, 2]. The technology was then introduced to other clinical fields and is now increasingly employed in clinical skin research. Compared to imaging modalities such as ultrasound and confocal microscopy, OCT has a resolution of up to 1 μm and penetration depth of around 1 mm which is adequate for the analysis of most skin diseases [3].

Figure 2 shows the OCT images captured from various anatomical sites, using SkinTell HD-OCT system (Agfa Healthcare, Mortsel, Belgium). The skin layer and structures are visible in the OCT images. Figure 3 (top) shows the OCT image of forearm, where the dermal-epidermal junction (DEJ) is clearly defined. Figure 3 (bottom) shows that the sweat duct can also be captured with OCT imaging modality.

Skin Layer Segmentation in OCT

Skin layer segmentation in OCT is a key step in skin analysis and has been one of our main works. To date, most methods and techniques developed in this area are machine learning based.

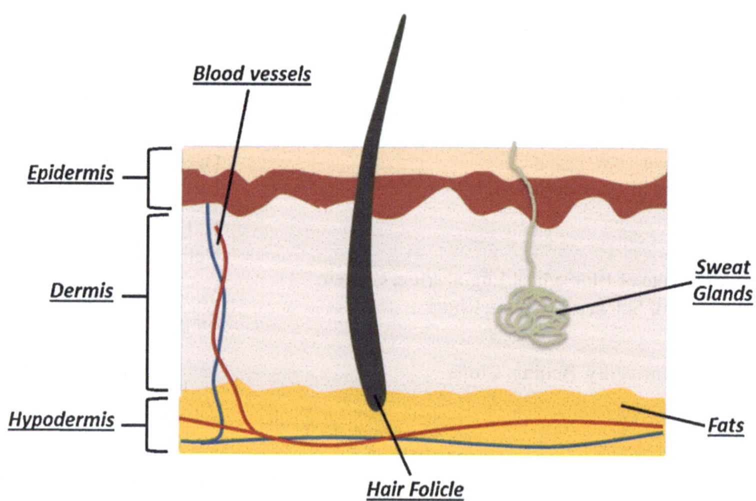

Fig. 1 Schematic diagram of skin layers and structures

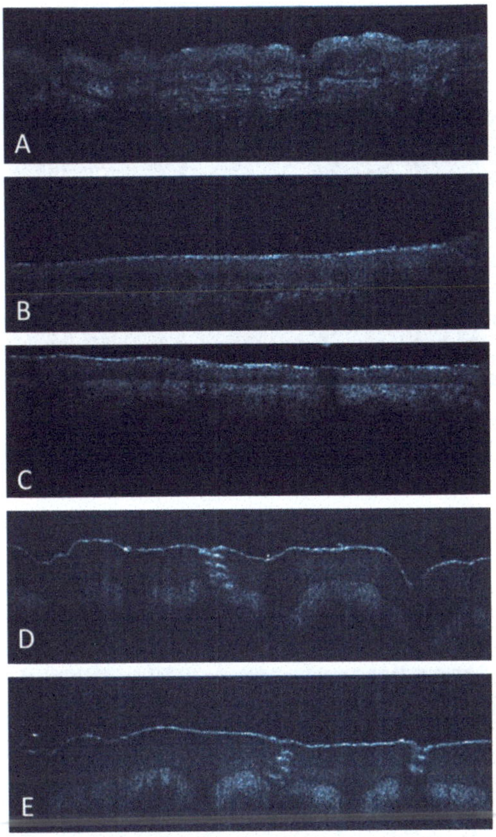

Fig. 2 High-definition OCT images of various anatomical sites: (**a**) Forearm, (**b**) Forehead, (**c**) Cheek, (**d**) Fingertip, (**e**) Palm

We previously proposed a method for skin layer segmentation [4]. In the study, we reported the use of a weighted least square (WLS)- based edge-preserving smoothing method to reduce speckle noise in OCT images, followed by using vertical intensity gradients to formulate the cost function which is used with a modified 2D graph-search algorithm to delineate skin surface in cross-sectional OCT images [4]. Next, the intersection between the epidermis and the dermis, known as the dermis-epidermis junction (DEJ), was detected by using the local integral projection as shown in Fig. 4.

The proposed method was evaluated on a dataset of 5 OCT volumes acquired from the SkinTell HD-OCT system (Agfa Healthcare, Mortsel, Belgium). Each volume is comprised of 512 cross-sectional 2D scans, with the size of each 2D scan as 200 × 640, at 3 μm voxel resolution. Comparison was made between epidermis segmentations performed on the original and pre-processed images. The average overlap ratio of 0.744 showed that the removal of speckle noise in OCT images was useful in skin layer segmentation. However due to reduced signal strength from intensity drop-off at the periphery of the volume as shown in Fig. 5 and the presence of hair which highly attenuates the signal from the underlying structures, the performance of DEJ detection was limited.

Subsequently, to avoid noticeable signal falloff, a method was proposed to perform ROI extraction to limit the segmentation to a centre 360 × 360 region as shown in Fig. 6 [5].

Using this modification with the 2D graph-based segmentation approach proposed earlier [4], the calculated skin surface roughness showed close adherence to human visual assessment in cross-sectional images with good signal quality which was only marginally affected by shadowing artefacts introduced by the presence of hair in the volume.

For the images with less visible epidermis, the segmentation accuracy is not as good as those with more visible epidermis. To counter this problem, we implemented a supervised 3D graph-based method that can give better segmentation accuracy in cases where the boundary is not clear [6]. ROI extraction and removal of speckle noise were performed and followed by the layer segmentation. As shown in Fig. 7, there are two phases in the segmentation process, namely training and testing.

In the training phase, training data were pre-processed to reduce noise. Thereafter, the intensity distributions around the boundary were modelled and used for computing the learned cost for the graph-based method. During the testing phase, test data underwent pre-processing and skin surface segmentation. Next, the learned cost was added to the gradient-based cost term to obtain a computed cost. Then, a max-flow/min-cut algorithm [7] is applied to detect the DEJ, thereby segmenting the epidermis. The proposed approach was evaluated on 10 OCT volumes which are captured from posterior

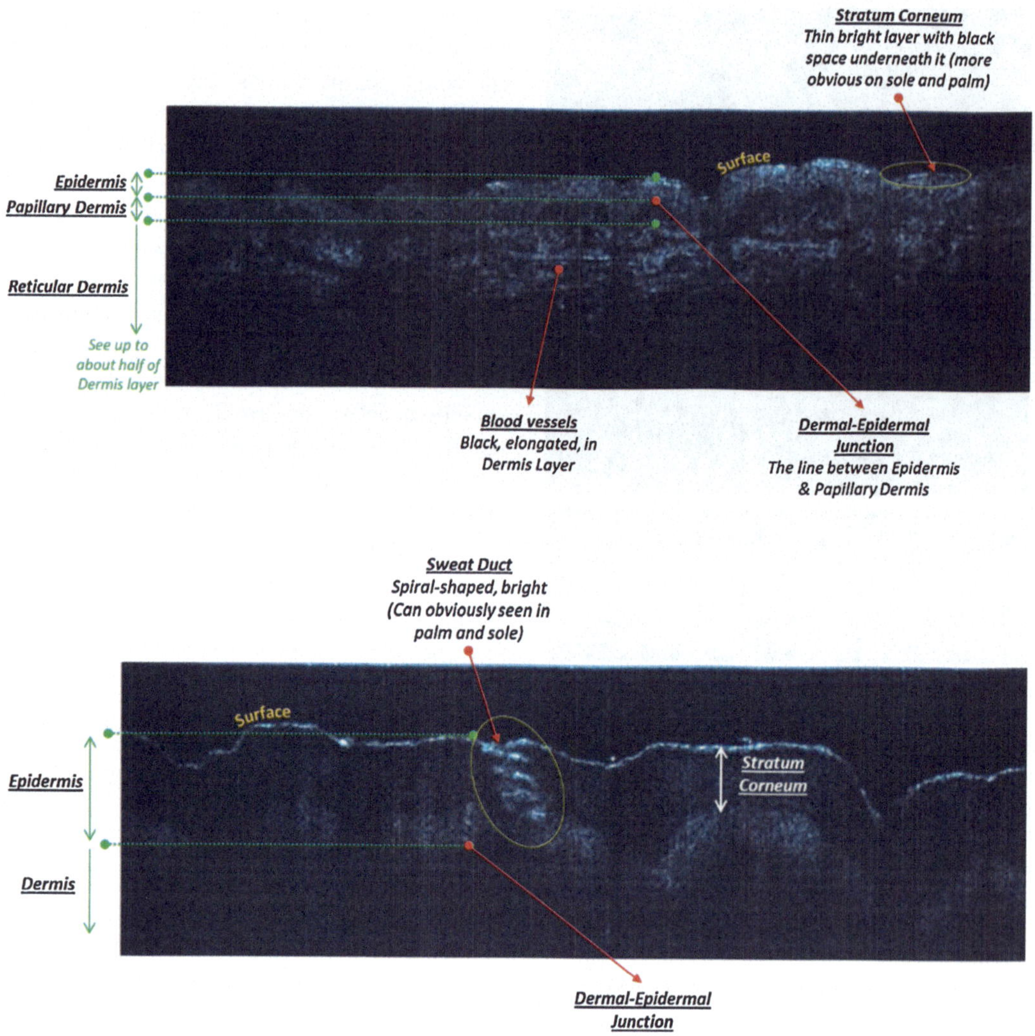

Fig. 3 Visible skin layers in OCT image of forearm (Top), sweat duct captured in OCT image of fingertip (Bottom)

and anterior forearm and faces of healthy subjects. The epidermis segmentation result was compared with two other related works: 2D graph-based OCT segmentation [4] and unsupervised segmentation [8]. The proposed method resulted the lowest unsigned error of 4.86 pixels as compared to 13.46 pixels and 5.35 pixels for the 2D and 3D methods, respectively. Figure 8 shows some of the sample segmentation results of the proposed method.

Although the proposed method was able to detect the regions with poorer visibility of the DEJ, images with shadowing due to hair remained a challenge as the abrupt signal drop-off beneath the hair did not conform to the modelled distribution, leading to inaccuracies in the detection of the boundary. Figure 9 shows sample images with shadowing due to hair.

To address the challenge caused by shadowing of hairs, a method as shown in Fig. 10 was proposed to segment the topmost layer in OCT images using 3D graphs with a novel cost function [9]. The cost function used both intensity gradients and uniformity of the regions surrounding the surface. With the context information across the neighbouring 2D scans, it ensured the smoothness of the surface across different scans and ignored any sudden loss of signal or a bump due to hair or skin scales.

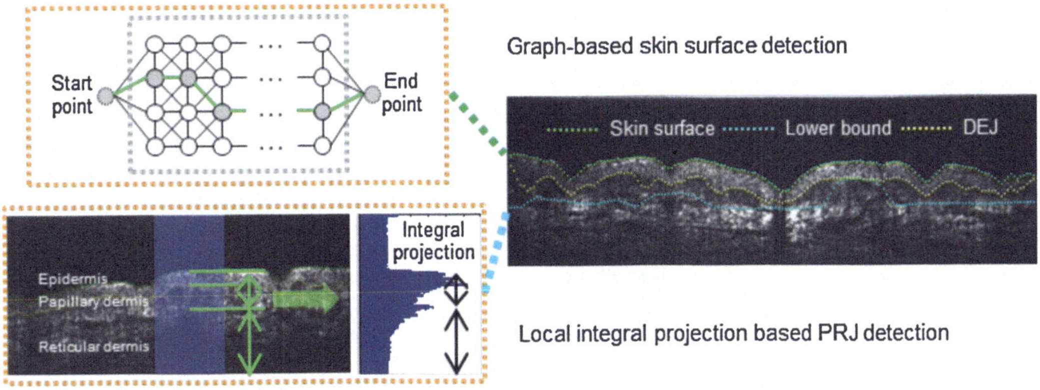

Fig. 4 Detected skin surface using modified 2D graph-search algorithm (green dotted line in right image); detected DEJ (yellow dotted line in right image); estimated lower bound of DEJ (blue dotted line in right image). Reproduction from Li et al. [4] with permission

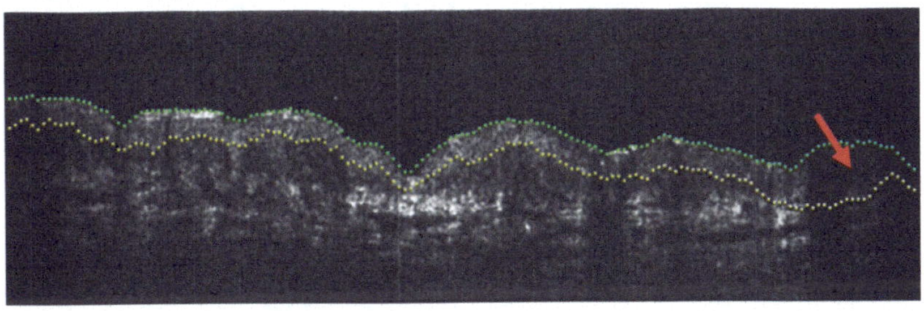

Fig. 5 Intensity drop-off at the right side the OCT image. Reproduction from Li et al. [4] with permission

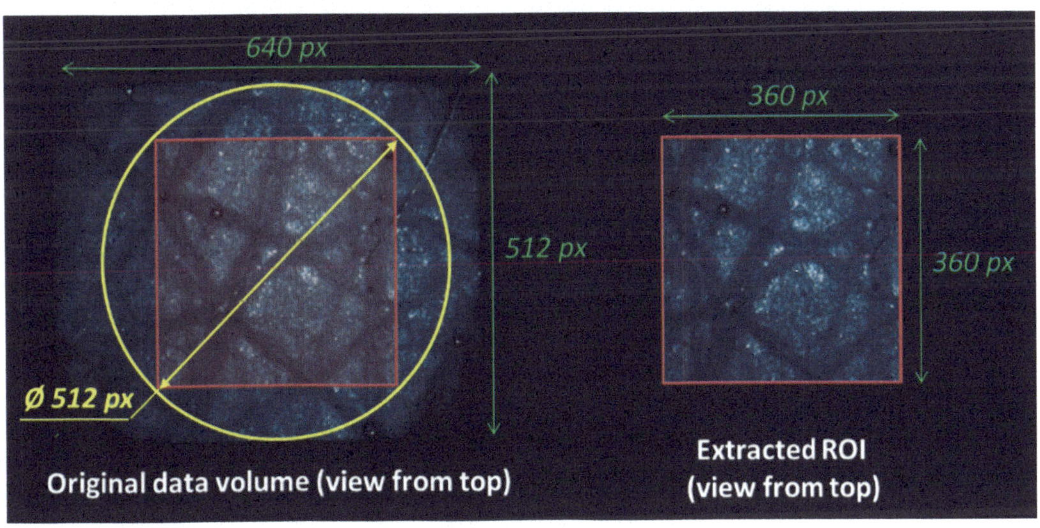

Fig. 6 Defining the ROI of skin OCT volume. Reproduction from Yow et al. [5] with permission

A total of 252 scans with shadowing effect were selected through visual inspection of 26 OCT volumes for evaluation of the accuracy of this proposed segmentation method and compared against the 2D-based approach, which did not consider the presence of hair [4] and a

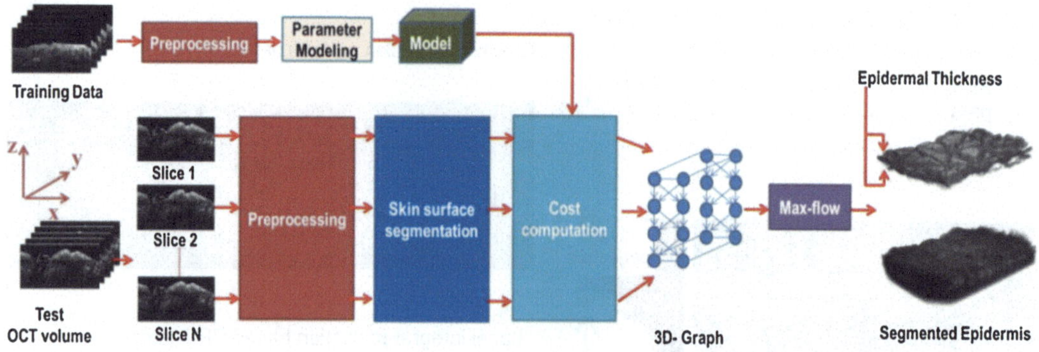

Fig. 7 Flowchart of proposed method for automated supervised 3D-graph segmentation. Reproduction from Srivastava et al. [6] with permission

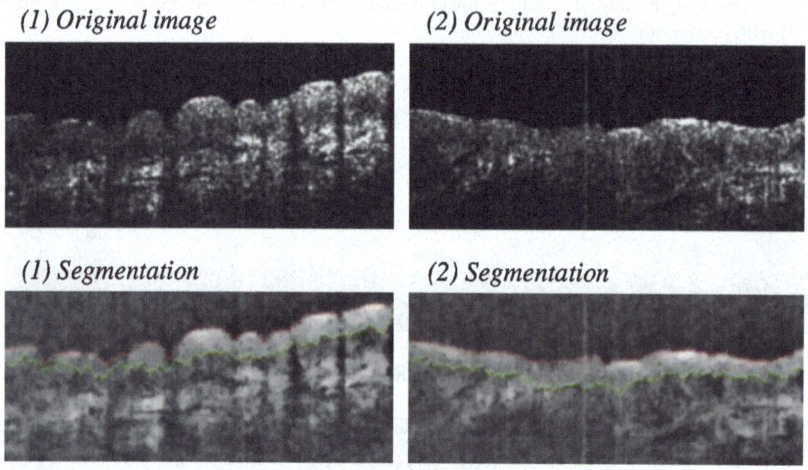

Fig. 8 Two example results of skin surface (red line) and epidermis segmentation (green line) using the supervised 3D-graph segmentation method. Reproduction from Srivastava et al. [6] with permission

Fig. 9 Four example cases where there is abrupt signal drop-off due to shadowing (yellow arrow). Reproduction from Srivastava et al. [9] with permission

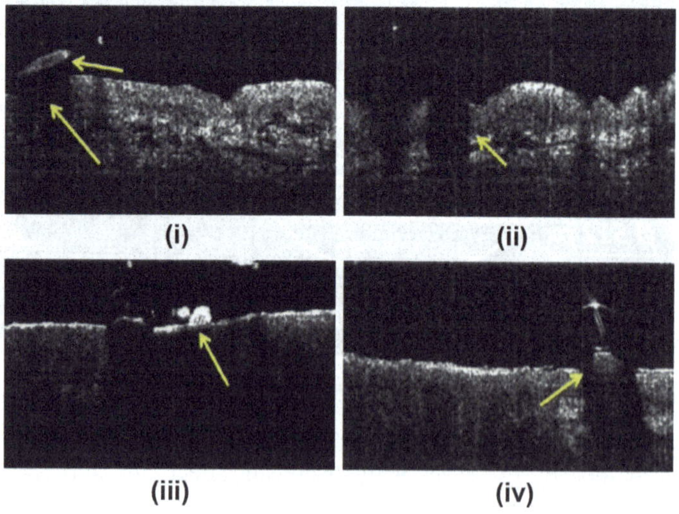

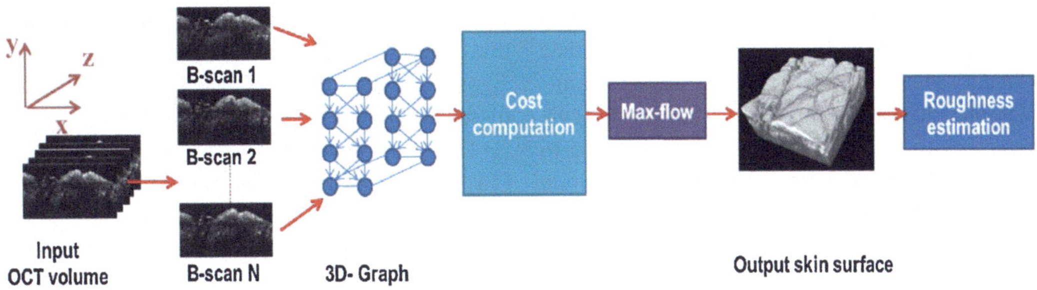

Fig. 10 Flowchart of the proposed approach for skin surface segmentation and roughness estimation. Reproduction from Srivastava et al. [9] with permission

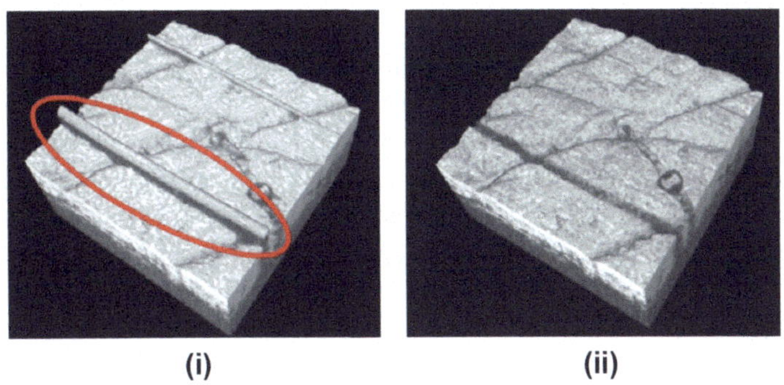

Fig. 11 Comparison of segmentation obtained using (i) 2D segmentation method and (ii) the proposed method. The encircled regions show strands of hair incorrectly segmented. Reproduction from Srivastava et al. [9] with permission

generic 3D graph-based surface segmentation approach [8]. Figure 11 compares the 3D visualization of the surface as segmented using the proposed method and the algorithm in [4]. As observed, the proposed method produces better segmentation being robust to the presence of hair.

The evaluation showed that the performance of the proposed method produced better results in particular with increased robustness to the presence of hair due to the use of context information from neighbouring cross-sectional scans. The proposed method produces a lower unsigned mean vertical error of 1.97 pixels, as compared to the two other works.

There are also other groups studied skin layer segmentation. One of the earlier works [10] used a novel shapelet-based image analysis technique to correlate an image with a kernel which is in the form of basic image shapes, then automatically identify the upper and lower boundaries of epidermis in OCT images of skin by using a column-wise and piece-wise line search algorithms. However, this technique fails to closely follow all the contours in wavy structures due to the shapelet kernel which is only processed in a single direction, normal to surface.

Another work on the segmentation of DEJ in OCT imaging first applied an automatic thresholding level and successive binary morphological openings and closing to help remove artefacts introduced from the applied gel between the probe and the tissue [11]. This was followed by an edge-following algorithm to trace the surface boundary, subsequently re-dressing the image to obtain a flat skin surface. The intensity profile for the image was computed and the first profile minima gave an estimation of the boundary position and then mapped to the OCT eight capture channels.

A more recent work [12] proposed a semi-automated approach by selecting start and end and arbitrary points in skin surface using an interactive framework by a graphical representation of an attenuation coefficient map through a uniform-cost search method and applied graph-based method to identify skin surface and the same procedure is applied to DEJ detection.

In another work [13] applied Gaussian and differential filter to emphasize the skin surface boundary and further refined by finding the minimal intensity values. Thereafter, the surface curvature is estimated and serves as a reference to produce a flatten OCT image. Real surface is then detected with differential filter procedure.

Applications: Roughness, ET

Skin layer segmentation is a crucial step in many dermatological applications and could potentially increase the clinical diagnostic capability and usability of skin OCT images. The topographic analysis of skin surface enables the determination of surface roughness which is useful in monitoring skin health and ageing. Furthermore, the identification and monitoring of epidermis is important in several clinical applications because the variation of epidermal thickness could depict signs of ageing and presence of dermatological diseases.

The team developed a skin analysis system [14] to automatically perform 2D segmentation and assess the skin surface topology and epidermal thickness. Skin furrows were extracted from the segmented skin surface and average furrows depth was computed by averaging the depth values at all extracted furrow point. Next epidermal thickness is defined by finding the average pixel difference between the segmented skin surface and DEJ at each column (Fig. 12).

The developed system is evaluated with 5 skin OCT volumes, which were selected from a pool of healthy forearm skin data. The team evaluated both skin topographic profile analysis and the dimensional properties of epidermis.

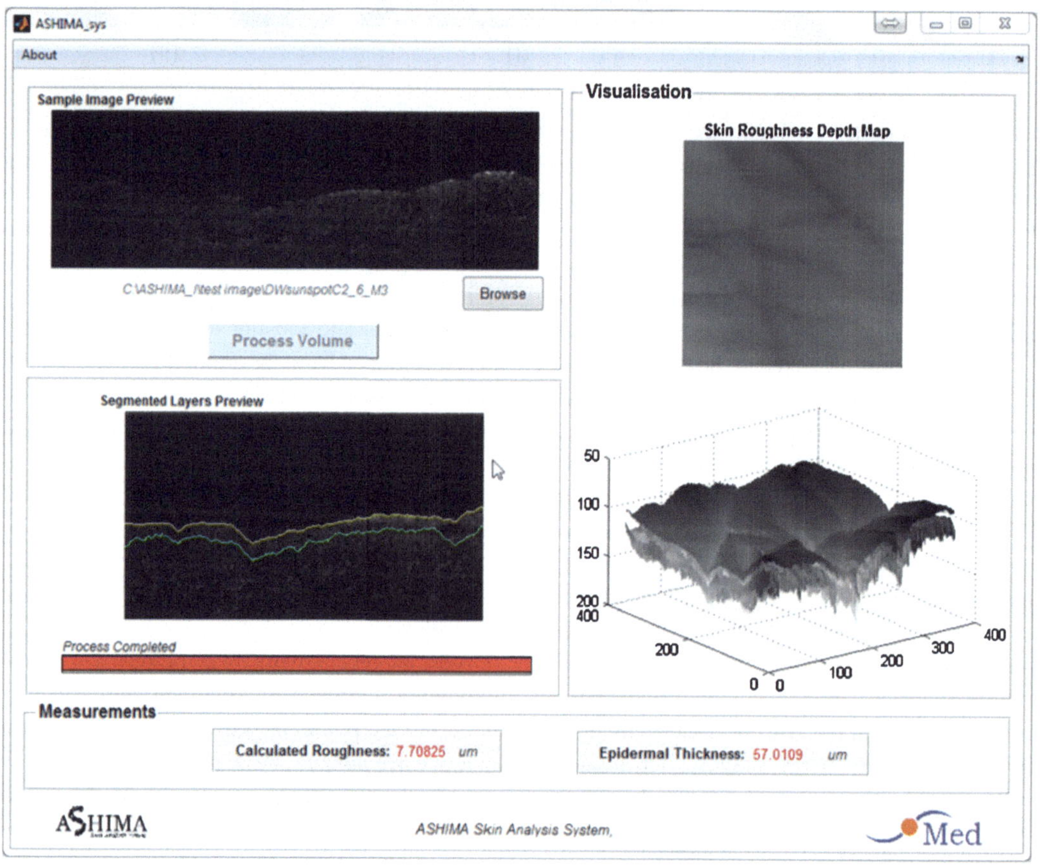

Fig. 12 Graphical user interface of skin OCT image analysis system. Reproduction from Yow et al. [14] with permission

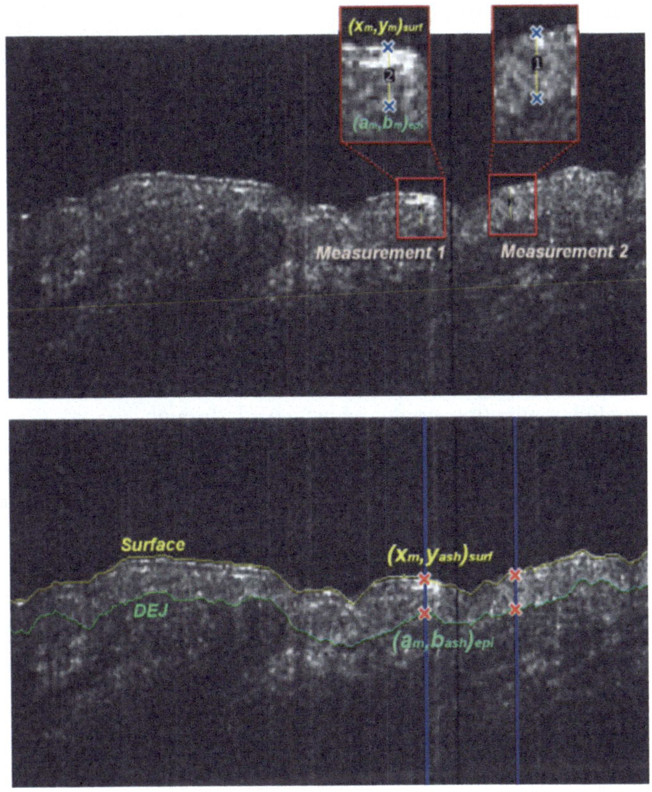

Fig. 13 (Top) Extracted middle slice of one of the OCT volumes with clinician's manual measurements; (Bottom) Measurement extracted from automated analysis. Reproduction from Yow et al. [14] with permission

Three segmented surface profiles were selected and presented in 3D-form for a senior consultant from National Skin Centre to visually assess and rank the images (1 as the least rough and 3 as the roughest). The computed furrow depth matches the manual visual appearance ranking.

In the evaluation of epidermal thickness measurements, the middle slice of the 5 volumes were extracted as JPEG files and manually measured thickness (between $(x_m,y_m)_{surf}$ and $(a_m,b_m)_{epi}$) at two distinctive locations in each image as shown in Fig. 13. These manual measurements are then compared to the automated measurements obtained. The average error across the entire dataset was 3.81 μm and there was no apparent difference if the measurement was applied in the image centre or at other locations which suggests that the automated system performs consistently across the image.

The team has also recently applied the 3D graph-based segmentation [9] to skin roughness estimation and shows a high correlation of at least 0.95 with manual assessment by 5 clinicians. The skin roughness estimation also takes into consideration for surfaces slanted at a significant angle from the x-axis by applying principal component analysis (PCA) to the points on the segmented surface.

In another work [13], skin surface roughness parameters based on ISO 25178 part 2 standard definitions were used to identify the effect of cosmetics on human skin. They compared the results with the PRIMOS device and showed that skin surface geometry acquired from 3D OCT images quantified complex wrinkle structure well.

Deep Convolutional Networks in Skin Imaging

In recent years, there has been an upsurge of interest in the application of deep convolutional networks for a variety of tasks. The rapid development of deep networks was motivated by the increasing availability and size of 'big' data and images which required more efficient architectures for tasks such as image classification. Features, which were previously manually designed, would instead be intrinsically extracted from a multitude of filters optimized directly from the data. This led to the development of AlexNet [15], which outperformed all other techniques at the 2012 ILSVRC (ImageNet Large Scale Visual Recognition Challenge) by at least 10%. The interest in deep networks also led to the realization that large scale parallel computing could be achieved through graphic processing units (GPU), motivating ever-increasing GPU capabilities which enabled even more complicated network architectures with improved performance. For an in-depth overview of the development of application of deep convolutional networks, the reader is urged to refer to recent review articles [16].

In medical imaging, the use of deep networks has been widely reported [17, 18], in fields ranging from radiology [19] and ophthalmology [20] to cardiology [21] and oncology [22]. In the following sections, we review recent applications of deep convolutional networks in dermatology.

Deep Learning for Classification of Dermoscopy Images

The most extensively reported application of deep learning in dermatology to date has been in the classification of dermoscopy images. A dermoscope is a hand-held device used for the inspection of skin lesions, primarily for the assessment of melanoma. In [23], the authors used Google's Inception v3 CNN architecture which had been pre-trained on the 2014 ImageNet Challenge (1.28 million images, 1000 classes). Transfer learning was used to adapt the pre-trained feature discriminators for the dermoscopy dataset, which consisted of 129,450 images with 2032 image-level classifications. In subsequent test of the trained network on a set of previously unseen, biopsy-confirmed images, the performance was shown to be on par with board-certified dermatologists in the identifications of the most common and the most serious skin cancers, achieving AUCs in excess of 0.95. A similar outcome was achieved in [24] when the performance of a CNN based on InceptionV4 was compared against 58 dermatologists, including skilled practitioners with more than 5 years of experience, however using a much smaller test set of 300 dermoscopy images. Other recent articles have described the use of different CNN architectures, such as ResNet [25] or the use of ensemble networks [26] for similar applications in dermoscopy images.

Deep Learning for Classification of Full Field OCT Images

The use of deep learning in OCT imaging for skin has been comparatively less well-reported. In Mandache et al. [27], a convolutional neural network was implemented for the detection of basal cell carcinoma regions in full field OCT (FF-OCT) images. FF-OCT is a coherence imaging technique in which *en face* images of ex vivo samples are generated with a greater imaging depth (200 μm) than confocal while achieving a lateral resolution of up to 1 μm. In this work, the authors manually annotated 40 FF-OCT images, which approximately 25% of each image were labelled as either normal or basal cell carcinoma (BCC)—a type of skin cancer, while the rest of the image was left unlabelled. Each FF-OCT image was split into patches of 256 × 256, and the patches were then used for training and testing. The authors customized a 10-layer network and achieved an overall classification accuracy of 95.93%.

Classification of Cross-Sectional OCT 2D Scans

While melanomas are visible on the surface of the skin and can be identified from the pigmentation of the affected skin area, basal cell carcinomas, which occur at the junction of the dermis and epidermis, are less apparent from dermoscopy images. This has been a major motivation in the development of dermatological OCT

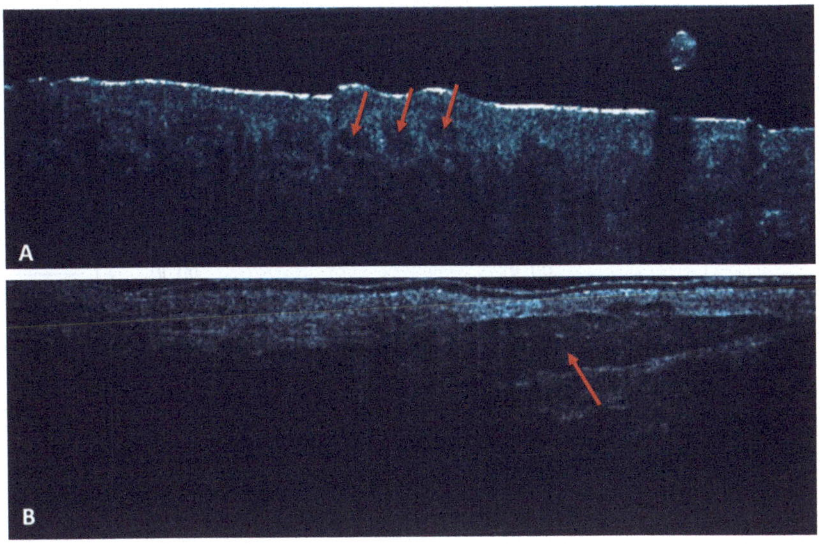

Fig. 14 BCC cells indicated by red arrows in skin OCT images: (**a**) Dark blobs, (**b**) Low intensity structures with epidermis layer being compressed

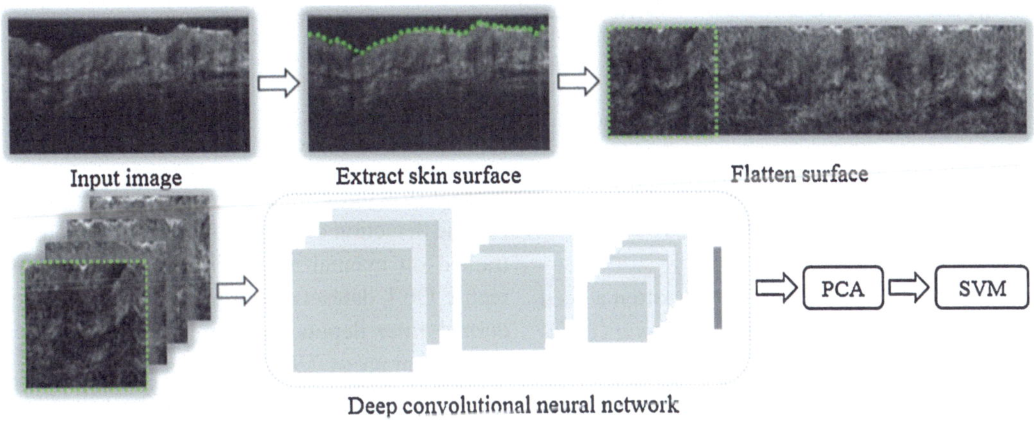

Fig. 15 Flowchart of proposed framework for automatic detection of BCC abnormalities in skin OCT images. Reproduction from Li et al. [29] with permission

systems which enable sub-epidermal visualization of the skin structure and any underlying lesions [28]. Figure 14 shows sample OCT images of BCC abnormalities appeared below epidermis layer.

In a proposed workflow, after acquisition of the OCT images, a dermatologist reviews the cross-sectional scans to identify suspicious regions suggestive of BCC. Li et al. proposed a framework as shown in Fig. 14 to automate the detection of BCC in cross-sectional OCT scans [29] (Fig. 15).

A graph-based method was first used to identify the skin surface, followed by the extraction of patches from the flattened surface. Four patches were generated each cross-sectional scan from a total of seven normal and seven BCC subjects. Pre-trained AlexNet, VGG-16, VGG-19, and GoogLeNet architectures were used to generate image features, followed by the use of a support vector machine for classification. From Fig. 16, the authors showed that the system using VGG-16-based image descriptors was the most optimal with an AUC of 0.935.

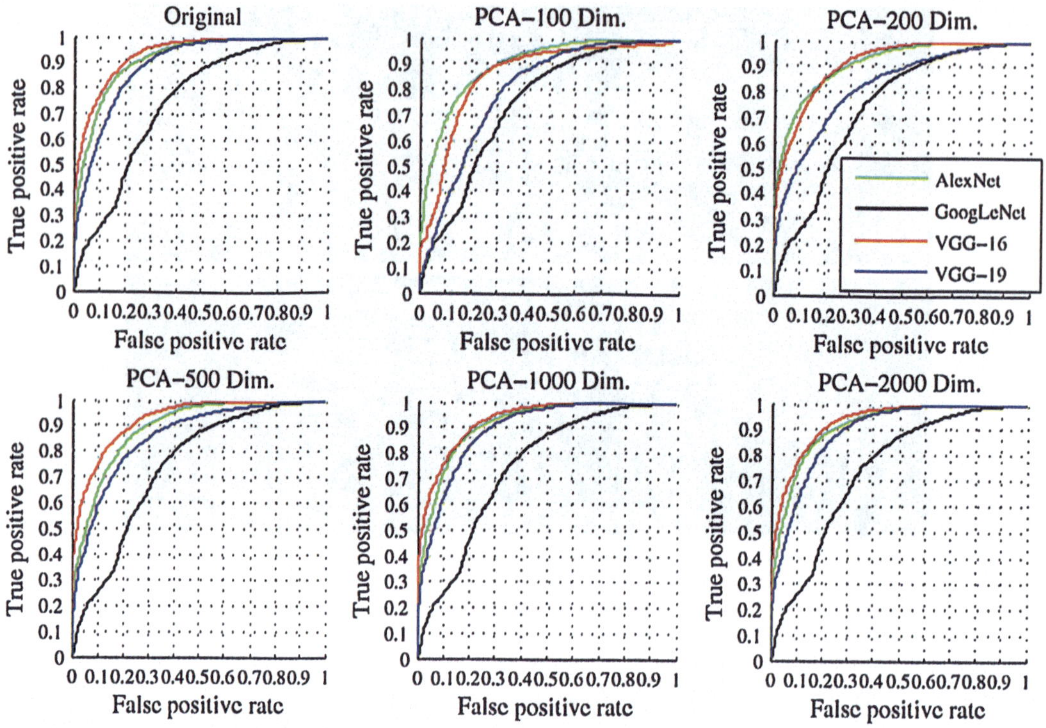

Fig. 16 BCC abnormalities detection results in ROC curves. Reproduction from Li et al. [29] with permission

Semantic Segmentation in Cross-Sectional OCT Images

In addition to classification, there is often a need to identify and localize structures or layers of interest in a cross-sectional OCT image. Recently, a system to automatically determine the volume of dermal fillers using convolutional neural network trained on volumetric images generated from a custom-built OCT was proposed [30]. As dermal fillers are absorbed by the surrounding tissue over time, there is interest to non-invasively assess the absorption rate of the filler material. In this work, the authors adapted the U-net architecture [31] for segmentation. Briefly, the U-net architecture adds a parallel up-sampling arm to the down-sampling arm typically used in image-based convolutional nets, resulting in the characteristic U-shaped architecture. Connections at steps between the arms provide high-level context to local features, and U-net-based systems have been successfully shown to be able to segment structures on biomedical images, including corneal structures in anterior segment OCT [32]. For the detection of dermal fillers, the U-net was trained on manual annotations from 100 volumetric OCT datasets recorded from 67 different dermal filler depots in 24 mice, resulting in a mean accuracy of 0.9938 and a Jaccard similarity coefficient of 0.879 from sixfold cross-validation.

Although distinct anatomical layers such as the strateum corneum, epidermis and dermis can often be identified from dermal OCT images, together with structures such as sweat ducts and hair follicles, the use of convolutional nets for the detection and semantic segmentation of these structures has not been widely reported compared to their use in the detection of structures from ophthalmic OCT imaging [33].

Challenges

As a data-driven approach, the main challenge in the development of systems based on deep convolutional network architectures was and remains the availability and quality of data, particularly for medical imaging. In general image classification, sources of data can include social media and public online repositories. Further, labels are often well-defined and specialized training is unnecessary to identify objects in daily living.

In contrast, datasets are often small in medical imaging. While there have been reports of large datasets such as in Esteva et al. [23] and Gulshan et al. [34], medical imaging datasets are usually several orders smaller than general image datasets. Further, with increasing awareness of privacy and regulatory requirements, sharing of data to augment dataset size is often a complicated process. Medical imaging datasets are often highly biased. Case-control datasets usually consist of subjects with late-stage or severe forms of the condition against normal subjects which is non-representative of use cases such as screening. While population-based datasets may be representative for screening, the limited numbers of positive disease cases in a population-based prevalence model tend to result in unbalanced datasets which can lead to bias in the trained model. Another challenge is in the labelling. In general image datasets, labels are often discrete, objective and obvious. In diseases, the labelling is often on a graduated scale ranging from normal to late-stage, and clinical grading of the severity of a disease from an image can be subjective.

Semantic annotation is also expensive, labour-intensive and subjective. In the annotation for semantic segmentation, objects have to be labelled and discriminated from other objects, sometimes at a pixel level. However, manual segmentation is highly time-consuming, in particular for volumetric scans, which can be up to hundreds of cross-sectional images for a single volume. Efficient methods for the annotation of volumetric structures have ranged from interpolating manual marking at intervals to using a preliminary or prior segmentation as an initial contour.

Finally, imaging the skin using OCT presents some unique challenges as well. Due to the opacity of the skin surface and heterogeneity of the skin, there is high scattering which leads to high signal falloff with increased imaging depth. This also affects the visibility of the different layers which can be difficult to discern. Further, as the largest organ in the human body, there are differences when imaging skin from different parts of the body, which necessitate the development of specific associated models, since the structural appearance of the skin can vary widely. This is in contrast to OCT imaging used in ophthalmic applications, where the region of tissue to be imaged is relatively fixed either posteriorly at the retina or anteriorly at the cornea and anterior segment of the eye. As such, ophthalmic OCT imaging protocols are relatively better established, with the use of OCT in ocular examinations becoming a standard clinical routine, having replaced colour photography in many practices. This also has implications for OCT image analysis, as consistent imaging protocols favour the development of image analysis tools for both screening and prognosis. Comparatively, the use of skin OCT currently largely remains situational, and clinical and imaging protocols for the use of skin OCT are still being developed.

Conclusions

Non-invasive imaging of the skin, which is the largest tissue in the human body, by optical coherence tomography facilitates investigation into underlying skin conditions which may not be apparent on the surface of the skin. The depth-resolved OCT imaging technique also allows micrometer-resolution quantification of skin structures and characteristics which could be useful in both clinical and cosmeceutical applications. However, the large volume of data generated in skin OCT requires automated methods for effective analysis. Currently, most approaches in the analysis of skin OCT volumetric data have relied on conventional analytic techniques. While deep convolutional networks have received a lot of attention and recently have been widely reported in medical imaging, the use of deep convolutional networks in dermatology is relatively limited compared to other fields such as ophthalmology. This is largely due to limited availability of skin OCT data and labels, which are the key challenges in the development of deep networks.

With rapid developments in OCT imaging techniques, decreasing costs, and increasing traction in clinical and cosmeceutical applications, there will be a greater need for advanced analytical approaches incorporating data-driven techniques such as deep networks to fully realize the potential of a non-invasive optical biopsy approach using OCT.

References

1. Fercher AF, Mengedoht K, Werner W (1988) Eye-length measurement by interferometry with partially coherent light. Opt Lett 13(3):186–188. https://doi.org/10.1364/OL.13.000186
2. Huang D, Swanson EA, Lin CP, Schuman JS, Stinson WG, Chang W, Hee MR, Flotte T, Gregory K, Puliafito CA et al (1991) Optical coherence tomography. Science (New York, NY) 254(5035):1178–1181
3. Mamalis A, Ho D, Jagdeo J (2015) Optical coherence tomography imaging of normal, chronologically aged, photoaged and photodamaged skin: a systematic review. Dermatol Surg 41(9):993–1005. https://doi.org/10.1097/dss.0000000000000457
4. Li A, Cheng J, Yow AP, Wall C, Wong DWK, Tey HL, Liu J (2015) Epidermal segmentation in high-definition optical coherence tomography. In: 2015 37th Annual International Conference of the IEEE Engineering in Medicine and Biology Society (EMBC), 25–29 Aug. 2015, pp 3045–3048. https://doi.org/10.1109/EMBC.2015.7319034
5. Yow AP, Cheng J, Li A, Wall C, Wong DWK, Liu J, Tey HL (2015) Skin surface topographic assessment using in vivo high-definition optical coherence tomography. In: 2015 10th International Conference on Information, Communications and Signal Processing (ICICS), 2–4 Dec. 2015, pp 1–4. https://doi.org/10.1109/ICICS.2015.7459853
6. Srivastava R, Yow AP, Cheng J, Wong DWK, Tey HL (2017) Supervised 3D graph-based automated epidermal thickness estimation. In: 2017 IEEE 2nd International Conference on Signal and Image Processing (ICSIP), 4–6 Aug. 2017, pp 297–301. https://doi.org/10.1109/SIPROCESS.2017.8124552
7. Boykov Y, Kolmogorov V (2004) An experimental comparison of min-cut/max-flow algorithms for energy minimization in vision. IEEE Trans Pattern Anal Mach Intell 26(9):1124–1137. https://doi.org/10.1109/tpami.2004.60
8. Kang L, Xiaodong W, Chen DZ, Sonka M (2006) Optimal surface segmentation in volumetric images-a graph-theoretic approach. IEEE Trans Pattern Anal Mach Intell 28(1):119–134. https://doi.org/10.1109/TPAMI.2006.19
9. Srivastava R, Yow AP, Cheng J, Wong DWK, Tey HL (2018) Three-dimensional graph-based skin layer segmentation in optical coherence tomography images for roughness estimation. Biomed Opt Express 9(8):3590–3606. https://doi.org/10.1364/BOE.9.003590
10. Weissman J, Hancewicz T, Kaplan P (2004) Optical coherence tomography of skin for measurement of epidermal thickness by shapelet-based image analysis. Opt Express 12(23):5760–5769. https://doi.org/10.1364/OPEX.12.005760
11. Josse G, George J, Black D (2011) Automatic measurement of epidermal thickness from optical coherence tomography images using a new algorithm. Skin Res Technol 17(3):314–319. https://doi.org/10.1111/j.1600-0846.2011.00499.x
12. Taghavikhalilbad A, Adabi S, Clayton A, Soltanizadeh H, Mehregan D, Avanaki MRN (2017) Semi-automated localization of dermal epidermal junction in optical coherence tomography images of skin. Appl Opt 56(11):3116–3121. https://doi.org/10.1364/AO.56.003116
13. Askaruly S, Ahn Y, Kim H, Vavilin A, Ban S, Kim PU, Kim S, Lee H, Jung W (2019) Quantitative evaluation of skin surface roughness using optical coherence tomography in vivo. IEEE J Sel Top Quantum Electron 25(1):1–8. https://doi.org/10.1109/JSTQE.2018.2873489
14. Yow AP, Cheng J, Li A, Srivastava R, Liu J, Wong DWK, Tey HL (2016) Automated in vivo 3D high-definition optical coherence tomography skin analysis system. In: 2016 38th Annual International Conference of the IEEE Engineering in Medicine and Biology Society (EMBC), 16–20 Aug. 2016, pp 3895–3898. https://doi.org/10.1109/EMBC.2016.7591579
15. Krizhevsky A, Sutskever I, Hinton GE (2012) ImageNet classification with deep convolutional neural networks, vol 25. https://doi.org/10.1145/3065386
16. Rawat W, Wang Z (2017) Deep convolutional neural networks for image classification: a comprehensive review. Neural Comput 29(9):2352–2449. https://doi.org/10.1162/NECO_a_00990
17. Greenspan H, Ginneken B, Summers RM (2016) Guest editorial deep learning in medical imaging: overview and future promise of an exciting new technique. IEEE Trans Med Imaging 35(5):1153–1159. https://doi.org/10.1109/TMI.2016.2553401
18. Litjens G, Kooi T, Bejnordi BE, Setio AAA, Ciompi F, Ghafoorian M, van der Laak JAWM, van Ginneken B, Sánchez CI (2017) A survey on deep learning in medical image analysis. Med Image Anal 42:60–88. https://doi.org/10.1016/j.media.2017.07.005
19. Choy G, Khalilzadeh O, Michalski M, Do S, Samir AE, Pianykh OS, Geis JR, Pandharipande PV, Brink JA, Dreyer KJ (2018) Current applications and future impact of machine learning in radiology. Radiology 288(2):318–328. https://doi.org/10.1148/radiol.2018171820

20. Ting DSW, Pasquale LR, Peng L, Campbell JP, Lee AY, Raman R, Tan GSW, Schmetterer L, Keane PA, Wong TY (2019) Artificial intelligence and deep learning in ophthalmology. Br J Ophthalmol 103(2):167–175. https://doi.org/10.1136/bjophthalmol-2018-313173
21. Shameer K, Johnson KW, Glicksberg BS, Dudley JT, Sengupta PP (2018) Machine learning in cardiovascular medicine: are we there yet? Heart 104(14):1156–1164. https://doi.org/10.1136/heartjnl-2017-311198
22. Levine AB, Schlosser C, Grewal J, Coope R, Jones SJM, Yip S (2019) Rise of the machines: advances in deep learning for cancer diagnosis. Trends Cancer 5(3):157–169. https://doi.org/10.1016/j.trecan.2019.02.002
23. Esteva A, Kuprel B, Novoa RA, Ko J, Swetter SM, Blau HM, Thrun S (2017) Dermatologist-level classification of skin cancer with deep neural networks. Nature 542:115. https://doi.org/10.1038/nature21056
24. Haenssle HA, Fink C, Schneiderbauer R, Toberer F, Buhl T, Blum A, Kalloo A, Hassen ABH, Thomas L, Enk A, Uhlmann L (2018) Man against machine: diagnostic performance of a deep learning convolutional neural network for dermoscopic melanoma recognition in comparison to 58 dermatologists. Ann Oncol 29(8):1836–1842. https://doi.org/10.1093/annonc/mdy166
25. Han SS, Kim MS, Lim W, Park GH, Park I, Chang SE (2018) Classification of the clinical images for benign and malignant cutaneous tumors using a deep learning algorithm. J Invest Dermatol 138(7):1529–1538. https://doi.org/10.1016/j.jid.2018.01.028
26. Marchetti MA, Codella NCF, Dusza SW, Gutman DA, Helba B, Kalloo A, Mishra N, Carrera C, Celebi ME, DeFazio JL, Jaimes N, Marghoob AA, Quigley E, Scope A, Yelamos O, Halpern AC (2018) Results of the 2016 International Skin Imaging Collaboration International Symposium on Biomedical Imaging challenge: comparison of the accuracy of computer algorithms to dermatologists for the diagnosis of melanoma from dermoscopic images. J Am Acad Dermatol 78(2):270–277.e271. https://doi.org/10.1016/j.jaad.2017.08.016
27. Mandache D, Dalimier E, Durkin JR, Boceara C, Olivo-Marin J, Meas-Yedid V (2018) Basal cell carcinoma detection in full field OCT images using convolutional neural networks. In: 2018 IEEE 15th International Symposium on Biomedical Imaging (ISBI 2018), 4–7 April 2018, pp 784–787. https://doi.org/10.1109/ISBI.2018.8363689
28. Boone M, Suppa M, Miyamoto M, Marneffe A, Jemec G, Del Marmol V (2016)In vivo assessment of optical properties of basal cell carcinoma and differentiation of BCC subtypes by high-definition optical coherence tomography. Biomed Opt Express 7(6):2269–2284. https://doi.org/10.1364/boe.7.002269
29. Li A, Cheng J, Yow AP, Srivastava R, Wong DW, Hong Liang T, Jiang L (2016) Automated basal cell carcinoma detection in high-definition optical coherence tomography. Conference Proceedings: Annual International Conference of the IEEE Engineering in Medicine and Biology Society IEEE Engineering in Medicine and Biology Society Annual Conference 2016:2885–2888. https://doi.org/10.1109/embc.2016.7591332
30. Pfister M, Schutzenberger K, Pfeiffenberger U, Messner A, Chen Z, Dos Santos VA, Puchner S, Garhofer G, Schmetterer L, Groschl M, Werkmeister RM (2019) Automated segmentation of dermal fillers in OCT images of mice using convolutional neural networks. Biomed Opt Express 10(3):1315–1328. https://doi.org/10.1364/boe.10.001315
31. Ronneberger O, Fischer P, Brox T (2015) U-Net: convolutional networks for biomedical image segmentation. In: Cham, Medical Image Computing and Computer-Assisted Intervention – MICCAI 2015. Springer International Publishing, pp 234–241
32. Santos VA, Schmetterer L, Stegmann H, Pfister M, Messner A, Schmidinger G, Garhofer G, Werkmeister RM (2019) CorneaNet: fast segmentation of cornea OCT scans of healthy and keratoconic eyes using deep learning. Biomed Opt Express 10(2):622–641. https://doi.org/10.1364/BOE.10.000622
33. De Fauw J, Ledsam JR, Romera-Paredes B, Nikolov S, Tomasev N, Blackwell S, Askham H, Glorot X, O'Donoghue B, Visentin D, van den Driessche G, Lakshminarayanan B, Meyer C, Mackinder F, Bouton S, Ayoub K, Chopra R, King D, Karthikesalingam A, Hughes CO, Raine R, Hughes J, Sim DA, Egan C, Tufail A, Montgomery H, Hassabis D, Rees G, Back T, Khaw PT, Suleyman M, Cornebise J, Keane PA, Ronneberger O (2018) Clinically applicable deep learning for diagnosis and referral in retinal disease. Nat Med 24(9):1342–1350. https://doi.org/10.1038/s41591-018-0107-6
34. Gulshan V, Peng L, Coram M, Stumpe MC, Wu D, Narayanaswamy A, Venugopalan S, Widner K, Madams T, Cuadros J, Kim R, Raman R, Nelson PC, Mega JL, Webster DR (2016) Development and validation of a deep learning algorithm for detection of diabetic retinopathy in retinal fundus photographs. JAMA 316(22):2402–2410. https://doi.org/10.1001/jama.2016.17216

Deep Learning Technique for Musculoskeletal Analysis

Naoki Kamiya

Abstract

Advancements in musculoskeletal analysis have been achieved by adopting deep learning technology in image recognition and analysis. Unlike musculoskeletal modeling based on computational anatomy, deep learning-based methods can obtain muscle information automatically. Through analysis of image features, both approaches can obtain muscle characteristics such as shape, volume, and area, and derive additional information by analyzing other image textures. In this chapter, we first discuss the necessity of musculoskeletal analysis and the required image processing technology. Then, the limitations of skeletal muscle recognition based on conventional handcrafted features are discussed, and developments in skeletal muscle recognition using machine learning and deep learning technology are described. Next, a technique for analyzing musculoskeletal systems using whole-body computed tomography (CT) images is shown. This study aims to achieve automatic recognition of skeletal muscles throughout the body and automatic classification of atrophic muscular disease using only image features, to demonstrate an application of whole-body musculoskeletal analysis driven by deep learning. Finally, we discuss future development of musculoskeletal analysis that effectively combines deep learning with handcrafted feature-based modeling techniques.

Keywords

Skeletal muscle · Musculoskeletal analysis · Musculoskeletal segmentation · Surface muscle · Deep muscle · Random forest · FCN-8s · 2D U-Net · 3D U-Net

Importance of Musculoskeletal Analysis and Skeletal Muscle Analysis

Musculoskeletal analysis is important in various situations. Muscles are divided into skeletal muscle, myocardium, and smooth muscle. In particular, the muscles that make up the heart are called the myocardium, while the muscles that make up the visceral organs are called the smooth muscles, and the smooth muscles have no muscle ganglion. Because skeletal muscle, which is the focus of

N. Kamiya (✉)
School of Information Science and Technology, Aichi Prefectural University, Nagakute, Japan
e-mail: n-kamiya@ist.aichi-pu.ac.jp

G. Lee, H. Fujita (eds.), *Deep Learning in Medical Image Analysis*, Advances in Experimental Medicine and Biology 1213,
https://doi.org/10.1007/978-3-030-33128-3_11

this chapter, adheres to bone and is directly involved in exercise, it is a fundamental focal point in musculoskeletal analysis.

In the field of orthopedics, analysis of bone itself is also important, and there are various bone analysis methods ranging from model-based segmentation to segmentation by deep learning [1–3]. We previously described the importance of skeletal muscle segmentation for orthopedic intervention and proposed model-based skeletal muscle recognition [4]. However, compared to the number of approaches that use deep learning for analysis of bone regions, there are far fewer approaches that use deep learning for skeletal muscle segmentation. Therefore, this chapter proposes a method of muscle recognition and analysis using deep learning.

Diseases related to skeletal muscle include myopathy, a myogenic disease, which is distinguished from neuropathy, a neurogenic disease; both diseases affect muscle function. Although most symptoms of myopathy involve atrophy, differentiating them from atrophy that occurs normally with aging is an important and difficult problem. Additionally, amyotrophic lateral sclerosis (ALS) and other similar afflictions focused on in this chapter are atrophic muscular diseases in which differential diagnoses do not exist; moreover, progress inhibitors have recently been approved by the U.S. Food and Drug Administration (FDA) [5], and treatment plans must appropriately differentiate muscular diseases with treatable atrophy. The measurement of skeletal muscle itself has become an important problem, because such measurements facilitate treatment of muscular diseases with atrophy; in turn, these treatments help patients live longer, healthier lives. Therefore, the measurement of skeletal muscle was addressed in a field study on the health and lifestyle habits of the elderly in Japan, which aimed to identify effective measures for improving their overall health [6].

Although the measurement of skeletal muscle quantity is required in musculoskeletal analysis as described above, accurate measurement of skeletal muscle quantity and automatic measurement of the muscle at each position are achievable for only a limited number of muscles, representing an unsolved problem in whole-body muscle analysis. Because skeletal muscle exists throughout the body, it is depicted in medical images obtained through various modalities. In tomographic CT images and MRIs, it will be difficult to search cross sections in which skeletal muscle is not depicted. However, CT images and tomographic MRIs that depict skeletal muscle photographed for the purpose of observing lesions can be utilized in support of automatic muscle analysis. In addition, because skeletal muscle is depicted in various cross sections of tomographic images, it is useful in automatic recognition of adjacent skeletal muscle as preprocessing for organ and lesion detection systems in computer-aided diagnosis (CAD) systems.

Musculoskeletal Recognition by Handcrafted Features and Its Limitations

Most skeletal muscle recognition methods that have been developed to date rely on handcrafted features. In the Computational Anatomy Project [7], we are performing computer-aided organ and disease recognition using CT images, where we adopted a method based on automatic recognition of normal structure. That is, similar to an analysis conducted by a doctor, by detecting the normal structure of the human body as a model in a computer (computational anatomy model), it is possible to detect an abnormality in an unknown case. Because the main purpose of this project was to construct a CAD system targeting the organ area, skeletal muscle was treated as one of the normal structures. In particular, in CT images, the density value distribution of the skeletal muscle overlaps with the density value distribution of the organ region, and thus it is difficult to segment the skeletal muscle and the other organ region using only the density value. Therefore, a muscle modeling technique that employs a shape model and probability model is commonly used [8]. In the Computational Anatomy Project, we have also achieved some success using computational models and muscle recognition for analysis of superficial and deep muscles [9]. In this study, we focused on the anatomical attachment point of the muscle, i.e., the origin and the insertion, and realized automatic recognition of muscles according to their position using a technique that arranged the shape model on the basis of the origin and the insertion. Automatic recognition of skeletal muscle was realized in surface and deep muscles of the thoracicoabdominal region. However, as described above, variations in skeletal muscle result not only from individual differences, but also from inter-individual differences due to aging, daily activities, and the characteristics of organs. Therefore, only a limited number of regions in the case database can be used for constructing a computational anatomical model of skeletal muscle with high recognition accuracy. In particular, there were only a limited number of scenarios in which shapes were comparatively clear or the boundaries between adjacent organs and adjoining muscle were clear.

In recent years, the Computational Anatomy Project has been developed into the Multi-disciplinary Computational Anatomy Project, with the purpose of achieving comprehensive understanding of the human body using medical image information over the four axes of space, time, function, and pathology [10]. Skeletal muscle recognition in the Multidisciplinary Computational Anatomy Project was designed with an emphasis on multi-axis awareness rather than computational anatomy. Muscle modeling ranges from micro to macro and considers the functional aspects of muscles [11]. In addition, in this multidimensional Computational Anatomy Project, we worked toward the automatic recognition and analysis of skeletal muscles with extended muscle regions and in contact with muscles and complex contour shapes [11]. In these computational anatomy modeling projects, nine regions (surface muscle: sternocleidomastoid muscle, trapezius muscle, supraspinous muscle, large pectoral muscle, intercostal muscle, oblique abdominal muscle, rectus abdominis muscle; deep muscle: psoas major muscle, iliac muscle) of skeletal muscle modeling-based segmentation were realized.

Figure 1 shows a conceptual diagram of skeletal muscle recognition by handcrafted feature modeling, which was realized during the Computational Anatomy Project and the Multidisciplinary Computational Anatomy Project. As Fig. 1 shows, we realized recognition and analysis of various muscles using torso CT images and whole-body CT images; however, the features of the muscles are one-dimensional points (landmarks; LM), two-dimensional running (muscle running), and three-dimensional shapes used to represent muscle features, i.e., the landmarks are acquired from the bone corresponding to the origin and insertion of the skeletal muscle, the running of the muscle is expressed by connecting LM on the bone, and the shape model based on gray values and probability distributions is arranged according to the running. In a technique based on handcrafted features, all the procedures from 1D to 3D worked sequentially until recognition according to the position of skeletal muscle was achieved. Therefore, it can be said that many dependent procedures must be performed to obtain initial information such as muscle quantity and intramuscular fat quantity, which are necessary for the site analysis of skeletal muscle. In general, the accuracy of each procedure greatly affects the accuracy of the final result.

As described above, when using handcrafted features in the recognition of skeletal muscle according to site, the 2D cross section or the shape of the muscle became a limited region with an easy-to-model shape. We then tackled recognition and analysis of the muscle using machine learning and deep learning techniques in the later stage of the multiple calculation anatomy project. The following sections describe recognition and analysis of muscles with more complicated shapes through deep learning technology.

Skeletal Muscle Segmentation Using Deep Learning

This section describes skeletal muscle segmentation using deep learning. As described previously, the problem of segmenting skeletal muscles in CT images is essentially the same as the problem of automatic recognition of organs. However, as

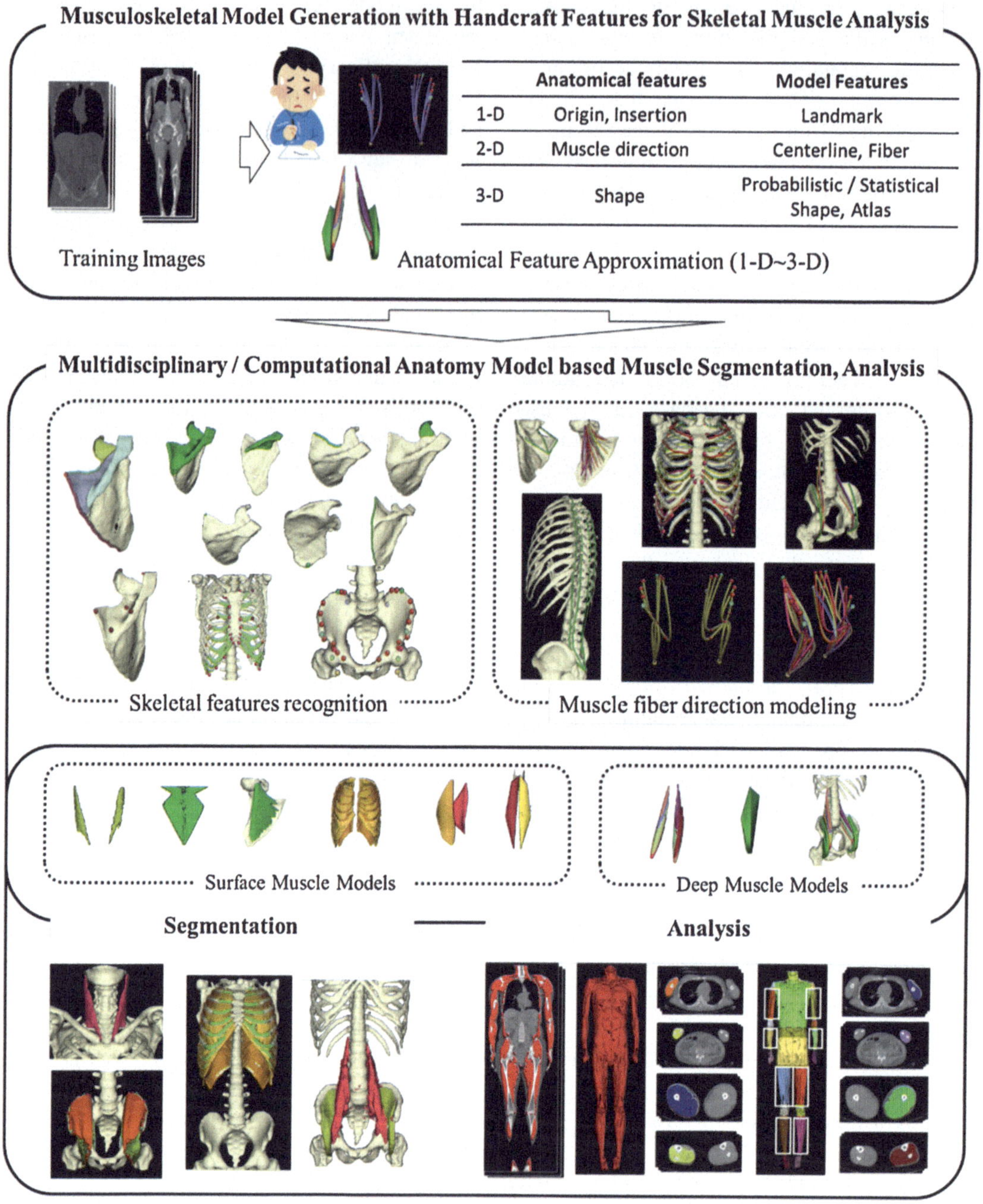

Fig. 1 Outline of the computational anatomy model for skeletal muscles analysis

described above, when the recognition of skeletal muscle by region is considered, the shape is complicated, and difficulties result from the juxtaposition of the organ region and skeletal muscle as well as the connection between skeletal muscles. In addition, very large individual differences in muscle mass cause difficulties in model-based skeletal muscle recognition. Prior to discussing skeletal muscle segmentation by deep learning, this section introduces the segmentation of erector spinae using random forest, the well-known machine learning algorithm.

The cross-sectional area of the erector spinae muscle is smaller in patients with COPD than in healthy individuals, and it has been found that this cross-sectional area indicates the prognosis of COPD patients [12]. However, the erector spinae is a very large group of muscles originating from the sacrum and located along the left and right vertebral columns. Therefore, manual measurement of the transverse area of the erector spinae is a time-consuming and error-prone task. We performed 3D recognition of the erector spinae muscle using the iterative random forest method and multiple sources of information [13]. Here, the original image and the probability map of the erector spinae, which is repeatedly refined, were used as sources, and improvements in recognition accuracy and high-speed segmentation were realized. During the learning process, three classifiers are trained. Classifier 1 is trained using low-resolution appearance features extracted from downsampled CT images. Then, the appearance features and probability map obtained by classifier 1 are combined to create trained classifier 2. Concurrently, the probability map obtained from classifier 1 is upsampled, and learned classifier 3 is obtained together with high-resolution appearance features. Thus, the final erector spinae segmentation results were obtained by combining classifiers 1 and 2, which are learned with low-resolution data, and classifier 3, which is trained with high-resolution data. Ten training cases were randomly selected from torso CT images from 20 cases. These were evaluated in 10 test cases, in which the average Dice coefficient (DC) was 93.0 ± 2.1% and the Jaccard similarity coefficient (JSC) was 87.0 ± 3.5%. Thus, highly accurate segmentation of the erector spinae muscle was achieved with 10 learning images and very limited data. Figure 2 shows the recognition result using proposed iterative random forest method. It was shown that machine learning-based segmentation according to the position of the skeletal muscle could be a robust technique for analyzing large and complex muscles for which shape models are difficult to generate.

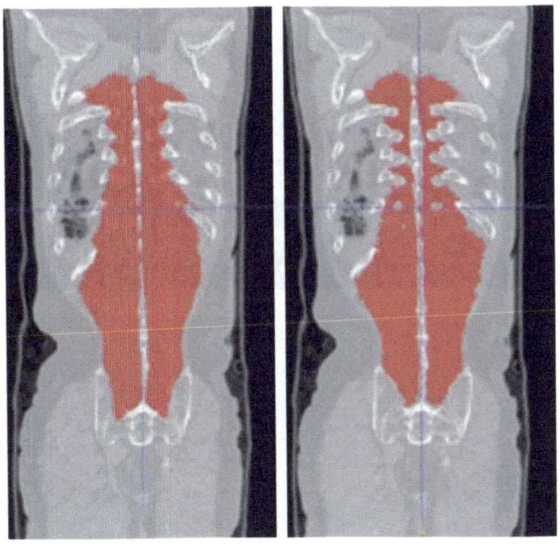

Fig. 2 Erector spinae muscle segmentation result using multi-scale iterative random forest method (left: recognition result, right: ground truth)

The following describes deep learning-based automatic recognition of the erector spinae in the cross sections of the 12 thoracic vertebrae. As stated above, the cross-sectional areas of the erector spinae muscle in the 12 thoracic sections indicate the prognosis of COPD patients. Therefore, we segmented the erector spinae muscle in the cross sections of the 12 thoracic vertebrae, and performed the segmentation of the muscle using deep learning and two-dimensional images [14]. The FCN-8s fully convolutional network (FCN) was used in order to utilize the results of the middle layer, in which detailed shape features are expected to be found. One-thousand correct images from 40 cases were prepared for learning of the erector spinae muscle in the cross sections of the 12 thoracic vertebrae; because the erector spinae muscle exists in the back, the learning was concentrated in the back half of the original image. Segmentation accuracy was evaluated using test images from 29 cases, and the average concordance rate of JSC was 82.4%. Figure 3 shows the segmentation of the erector spinae muscle in the cross section of the 12th thoracic vertebra. From top to bottom, the figure shows the original image from the top, the recognition result obtained by the model-based method, and the recognition result obtained by FCN-8s. Blue indicates a correct result, yellow indicates over-extraction, and red indicates unextracted areas.

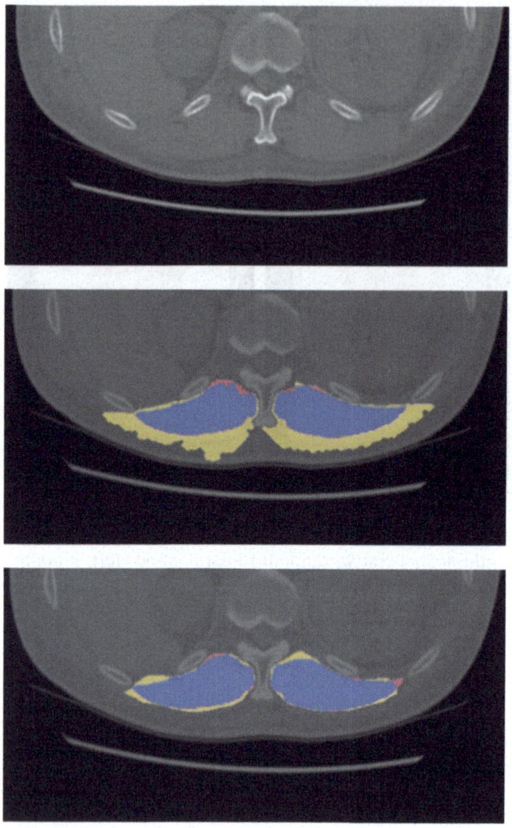

Fig. 3 Erector spinae muscle segmentation at the 12 thoracic vertebrae section using deep learning in cases with unclear skeletal muscle boundaries (upper: original CT, middle: model-based method, lower: deep CNN-based method (blue: matched, red: unextracted, yellow: over-extracted))

Because the boundary with the latissimus dorsi muscle is unclear, the conventional model-based method causes over-extraction during region expansion. With deep learning, it is proven that the target region is recognized even in these boundary areas.

Through deep learning of 2D images as shown above, automatic segmentation of skeletal muscle in a two-dimensional section could distinguish the boundary with the latissimus dorsi muscle, which could not be distinguished in conventional muscle modeling. Three-dimensional recognition of erector spinae by deep learning was carried out in a subsequent experiment.

In order to recognize the erector spinae as a volume, three-dimensional deep learning was used. This method can obtain three-dimensional recognition results using plural two-dimensional cross sections; thus, it is referred to as a 2.5-dimensional method. Here, FCN-8s, which obtained good results in the simultaneous segmentation of multiple organs in torso CT images [15], was used [16]. Various architectures have recently been proposed for medical image segmentation based on machine learning. In particular, U-net is well known for its efficacy in the field of medical image segmentation, and features a decoder employing an architecture similar to that of its encoder. Our group achieved automatic segmentation of multiple organs from torso CT images using segmentation based on FCN and the voting principle [17]. In a comparison of segmentation methods based on FCN and U-net, we obtained results showing that FCN and voting-based methods are realistic methods for segmenting CT images [17]. In the 2.5-dimensional FCN, three anatomical sections are input, and the simultaneous probability is calculated from the recognition result in each section; the label value with the highest simultaneous probability is selected as the result. By using the simultaneous probability, the recognition result of each cross section is judged comprehensively. The average DC and average JC of the erector spinae recognition results were $89.9 \pm 2.0\%$ and $81.7 \pm 3.2\%$, respectively, when evaluated by the leave-one-out method using the trunk CT images from 11 cases. Figure 4 shows the results of recognition of the erector spinae muscle using deep learning. In 10 cases similar to the three-dimensional recognition of the erector spinae muscle achieved by the random forest method described above, less segmentation was observed compared with the original random forest method; however, segmentation was successful, and boundaries with other skeletal muscles (such as the latissimus dorsi muscle) were recognized. Such boundaries were not recognized by the model-based method.

The results show that robust segmentation can be achieved with only 10 learning images. Skeletal muscle segmentation using deep learning is an effective technique that improves upon the manual 2D cross-sectional area measurements intended as an alternative measuring method to ease

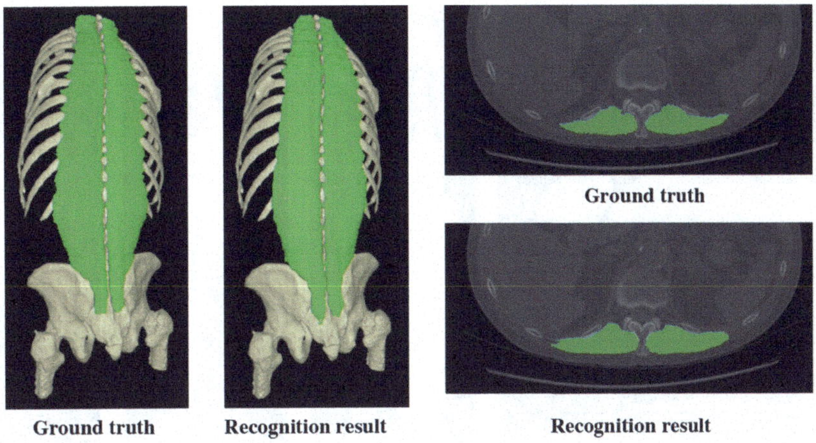

Fig. 4 Segmentation results of the erector spinae muscles using deep CNN

workloads incurred during automated 3D analysis. In particular, because the technique does not depend on the additional processing steps mentioned in the previous section, it can be said that high-quality segmentation results can be expected if a good training image is prepared. Moreover, high-quality segmentation results can be obtained in such a small number of cases, it is considered useful for efficient preparation of learning images needed for creating correct images used in developing deep learning-based segmentation algorithms.

Whole-Body Muscle Analysis Using Deep Learning

This section describes whole-body analysis of skeletal muscle using deep learning. As described above, the basic requirement for the analysis of skeletal muscle is the quantitative and automatic measurement of muscle mass, which is measured manually in current clinical situations. However, as previously mentioned, the technology for fully automated segmental analysis of skeletal muscle is still under development. Additionally, differentiating between diseases of the muscle itself and the quality of the muscle appears to be dependent on image analyses produced by the computer.

We have been working on whole-body muscle analysis using whole-body CT images. In particular, for whole-body CT images taken for diagnosis of amyotrophic lateral sclerosis (ALS), whole-body skeletal muscle recognition can be achieved using the active balloon model and skeletal muscle model [18]. We conducted muscle analysis [18] using texture analysis employing the Haralick's features of the region. However, the recognition of skeletal muscle over the whole body differs significantly for individual body types, and highly accurate surface muscle recognition has not been realized yet. Therefore, the area used for research on discrimination of myopathy was limited to limbs in which whole-body skeletal muscle [18] could be recognized. In this section, we first introduce a method to automatically classify atrophic myopathy through deep learning using the upper extremities, and then introduce whole-body musculoskeletal segmentation by deep learning.

First, the automatic classification of atrophic muscular disease in the upper arm and lower arm using deep learning is shown. Here, as an initial challenge, the automatic classification of ALS, which is a neurogenic disease, and myopathy, which is myogenic atrophy, was carried out [19]. It is an important problem to separate ALS, which is an intractable disease for which a therapy has not been established, from other atrophic diseases for which treatment is possible. Moreover, it is important to test the deep learning approach for diseases such as ALS, which can only receive exclusion diagnoses. Here, the architecture of

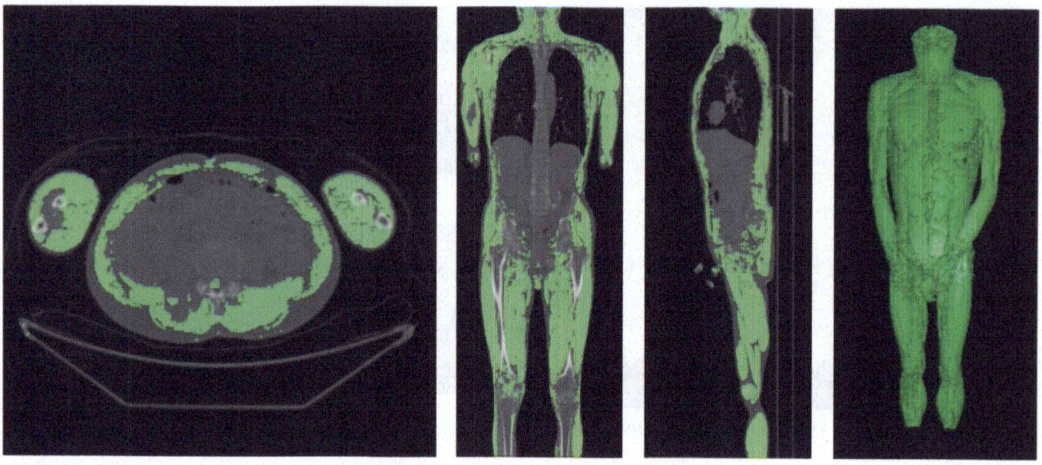

Fig. 5 Segmentation results of whole-body skeletal muscle using 2D U-Net

ResNet-50 was used. Drawing from five ALS cases and five myopathy cases, training images of 1678 upper arms and 1150 lower arms were prepared, and 171 upper and 130 lower arms were tested. The images were classified into ALS and myopathy. As a result, we obtained an average classification accuracy of 90.3% on the right forearm. This demonstrated the possibility of classifying diseases through deep learning with images of atrophic diseases, even when the differences were not recognized through visual observation. However, in order to fully diagnose ALS, it is necessary to evaluate more cases while considering the stage and type of the disease. In addition, methods for distinguishing the type of muscular atrophy and for analyzing the muscle region in more detail are necessary. This technique uses the results, which are roughly divided into 22 surface layer muscle regions [18] obtained by recognizing the body cavity through the active balloon model for whole-body skeletal muscle using conventional handcrafted characteristics, and differentiating the body cavity region from the whole body [18].

Using the above problems as a framework, we have been working toward fully automatic recognition of surface muscles in whole-body skeletal muscles [20]. As a preliminary experiment, the axial cross section was learned and recognized using 2D U-Net. When 50 cases were divided into training, test, and validation cases at a ratio of 8:1:1 according to the hold-out method, the segmentation results showed that DC was $81.7 \pm 0.9\%$ on average in three experiments. Figure 5 shows the results of whole-body skeletal muscle segmentation using 2D U-Net. Although only the axial CT slice was input, continuous recognition of surface muscle in the sagittal and coronal sections could be realized.

Similarly, bone segmentation was performed using 2D U-Net [21]. A related study has segmented bone from low-dose whole-body CT images using 2.5D U-Net [22]. Therefore, it is worth studying bone segmentation on whole-body plain CTs. The experiments were carried out using 17 whole-body CT images without contrast. A dataset consists of 12 training cases, two validation cases, and three test cases. We used whole-body CT images without contrast and trained 2-D U-Net with axial slices. The average dice coefficient of bone segmentation was 0.899, and this method was robust with regard to the position of hands. Figure 6 shows the segmentation result that bone segmentation by U-Net using only axial CT slices provides high performance with plain CT images. In the future, we will use this method for muscle segmentation to conduct bone classification using segmented bone images.

The whole-body musculoskeletal analysis presented in this section is still preliminary. However, deep learning has shown sufficient

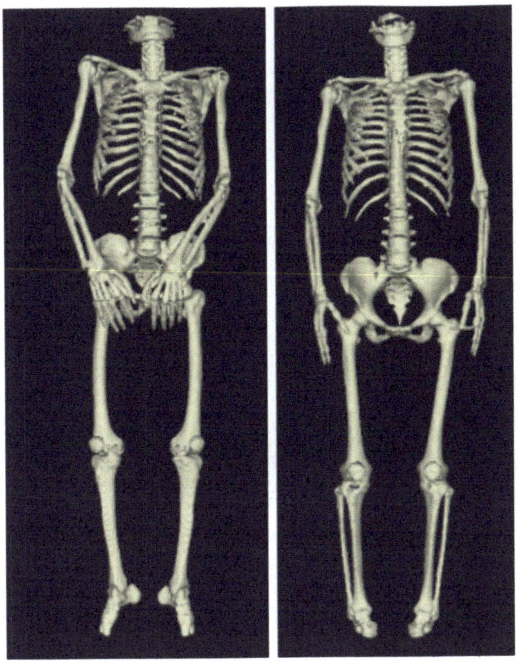

Fig. 6 Bone segmentation results in whole-body CT images using 2D U-Net

potential for classification and segmentation tasks, which are two major aspects of musculoskeletal muscle analysis. In the following section, we discuss the techniques needed to further advance whole-body musculoskeletal analysis.

Fusion of Deep Learning and Handcrafted Features in Skeletal Muscle Modeling

In the previous section, musculoskeletal segmentation and analysis using deep learning were described, and the possibility of musculoskeletal analysis by deep learning was shown.

In order to realize further advancements in musculoskeletal analysis in the future, deep learning must be effectively combined with the handcrafted features-based skeletal muscle modeling described at the beginning of this chapter. This is because correct images are required for deep learning. However, significant amounts of time are needed to paint all skeletal muscles and obtain a sufficient number of learning images, and the degree of difficulty in producing correct images is higher than for other organ regions. This section describes two approaches that aim to address these problems.

The first approach involves the simultaneous and automatic recognition of skeletal muscle, as well as the origin and stop position of skeletal muscle. Information about muscle and its adhesion position becomes important when conducting muscle analysis and muscle recognition while considering positions specified by handcrafted features. In particular, model-based methods provide information on the placement of muscle models. We carried out the simultaneous recognition of muscle and bone attachment positions on the muscle through deep learning. The 2.5-dimensional FCN employed in the automatic recognition of erector spinae was used. The mean DC of the erector spinae was $89.9 \pm 2.0\%$ and the mean JC was $81.7 \pm 3.2\%$ in 11 cases examined by the leave-one-out method. The average DC of the recognition results of the attachment area on the skeleton was $65.5 \pm 3.3\%$ and the average JC was $48.8 \pm 3.7\%$ [16]. Figure 7 shows the results of recognition of the attachment site between the erector spinae and the bone on the erector spinae. Because the bone attachment site area is small compared with the skeletal muscle area, the recognition accuracy is low; nevertheless, the attachment site of the muscle and bone is captured. Therefore, it is expected that the model can be applied to the analysis of muscle travel and the relation between muscle and disease.

Next, the challenge of classifying muscle groups obtained by deep learning through modeling the shape and running of muscle bundles [23] is shown. As described above, it is very difficult to create correct images from a sufficient number of cases for deep learning by separately painting muscle groups composed of multiple muscles. The results for the erector spinae muscle are shown here. Because the erector spinae muscle is composed of multiple muscles, it is labor intensive to prepare a learning image of the whole erector spinae muscle. We proposed a method to bundle three of the erector

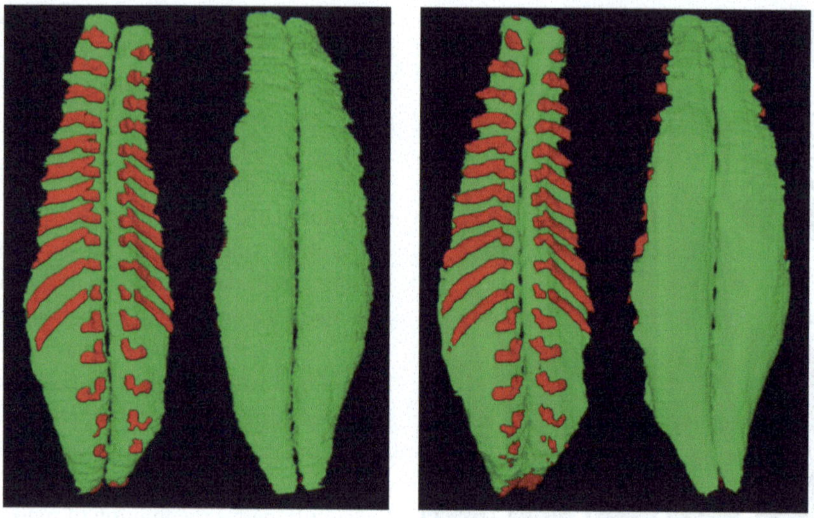

Fig. 7 Recognition results with origin and insertion on the erector spinae muscle

spinae muscles automatically recognized by deep learning (iliocostalis lumborum, iliocostalis thoracis, and longissimus thoracis); these particular muscles were selected because they are relatively distant from each other. The muscle bundle model is an ellipsoid connecting the beginning and end of each muscle, and the thickness of the ellipsoid is determined from the learning case. The erector spinae muscle can be segmented into three muscles by constructing a muscle bundle model containing the three muscles of the erector spinae muscle and arranging them into the recognized result of the erector spinae muscle using deep learning. Figure 8 shows the result of dividing the erector spinae muscle recognized by deep learning into three regions using the muscle bundle model. Green represents the erector spinae, yellow the iliocostalis lumborum, blue the longissimus thoracis, and orange the iliocostalis thoracis. Each modeled muscle bundle is convexly encapsulated and subdivided to the lower right. Because the erector spinae is composed of nine muscles, modeling of other muscles is also necessary. However, when we evaluated the ratio of the volume of each muscle bundle model to that of the erector spinae region in the convex hull region, the average Jaccard coefficient was 48.7 ± 6.8% in the right-hand region and 53.2 ± 4.2% in the left-hand region. However, as shown in the two-dimensional cross sections in Fig. 8, the positions of the muscles constituting the muscle group are shown, and for skeletal muscles for which preparing a correct individual muscle image is difficult, the recognition of the muscle through deep learning can be said to show the possibility of dividing muscles into regions according to position.

This section described the development of a new muscle analysis method that combines deep learning and muscle models. In the future, finer whole-body muscle analysis can be achieved by effectively using deep learning and modeling in conjunction with conventional handcrafted feature.

Conclusion

In this chapter, we described segmentation, recognition, and classification techniques we have developed to conduct musculoskeletal analysis using deep learning. Because musculoskeletal areas are large compared to areas occupied by organs, individual and intra-individual differences are also large, and it cannot be said that muscu-

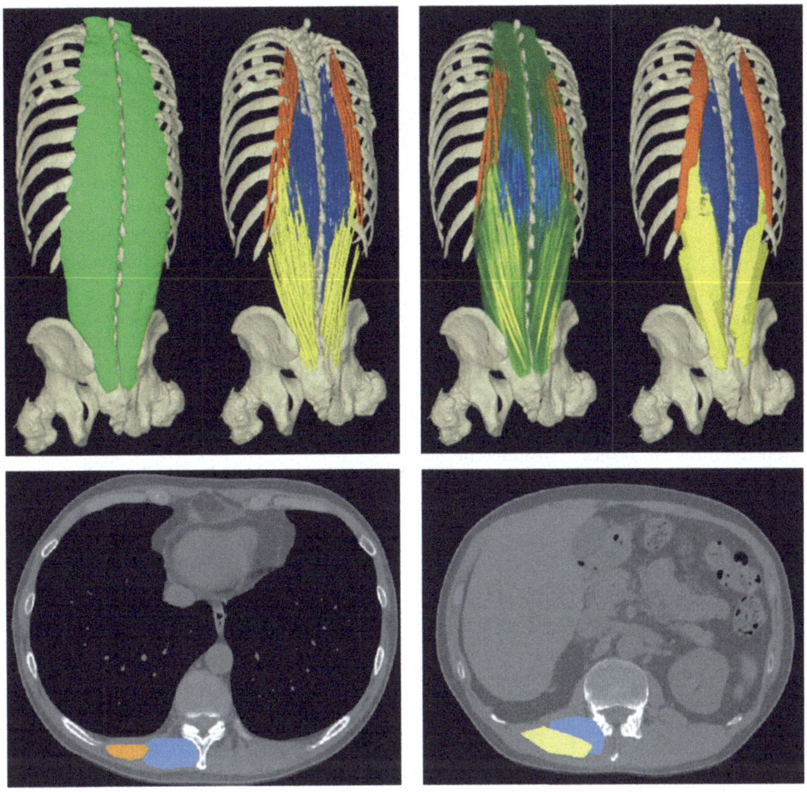

Fig. 8 Dividing the erector spinae muscle recognized by deep learning into three regions using the muscle bundle model (yellow: iliocostalis lumborum, blue: longissimus thoracis, orange: iliocostalis thoracis, green: erector spinae)

loskeletal analysis will be overwhelmingly simplified by the introduction of deep learning; however, handcrafted features also help improve the developed techniques. There is no doubt that deep learning techniques complement musculoskeletal analysis. As described in the latter half of the chapter, it is considered that musculoskeletal analysis can be markedly improved by effectively combining deep learning-based methods and handcrafted feature-based methods.

Acknowledgments The authors would like to thank all the members of the Fujita Laboratory in the Graduate School of Medicine, Gifu University, for their collaboration. We especially thank Ms. Oshima and Mr. Wakamatsu in the School of Information Science and Technology, Aichi Prefectural University for providing binary code and testing systems. This work was supported in part by a JSPS Grant-in-Aid for Scientific Research on Innovative Areas (Multidisciplinary Computational Anatomy, #26108005 and #17H05301) and for Challenging Exploratory Research (#16K15346), Japan.

References

1. Yu W, Liu W, Tan L et al (2018) Multi-object model-based multi-atlas segmentation constrained grid cut for automatic segmentation of lumbar vertebrae from CT images, intelligent orthopaedics. Adv Exp Med Biol 1093:65–71. https://doi.org/10.1007/978-981-13-1396-7_5
2. Zeng G, Zheng G (2018) Deep learning-based automatic segmentation of the proximal femur from MR images, intelligent orthopaedics. Adv Exp Med Biol 1093:73–79. https://doi.org/10.1007/978-981-13-1396-7_6
3. Yu W, Zheng G (2018) Atlas-based 3D intensity volume reconstruction from 2D long leg standing X-rays: application to hard and soft tissues in lower extremity, intelligent orthopaedics. Adv Exp Med Biol 1093:105–112. https://doi.org/10.1007/978-981-13-1396-7_9
4. Kamiya N (2018) Muscle segmentation for orthopedic interventions, intelligent orthopaedics. Adv Exp Med Biol 1093:81–91. https://doi.org/10.1007/978-981-13-1396-7_7

5. Rothstein JD (2017) Edaravone: a new drug approved for ALS. Cell 171(4):725
6. Ministry of Health, Labour and Welfare, JAPAN, National Health Promotion Movement in the 21st Century (Healthy Japan 21)
7. Kobatake H, Masutani Y et al (2017) Computational anatomy based on whole body imaging: basic principles of computer-assisted diagnosis and therapy. Springer
8. Hanaoka S, Kamiya N, Sato Y et al (2017) Skeletal muscle, understanding medical images based on computational anatomy models. Springer, pp 165–171
9. Fujita H, Hara T, Zhou X et al (2014) Model construction for computational anatomy: progress overview FY2009-FY2013. In: Proceedings of the Fifth International Symposium on the Project "Computational Anatomy", pp 25–35
10. Multidisciplinary Computational Anatomy and Its Application to Highly Intelligent Diagnosis and Therapy. http://wiki.tagen-compana.org
11. Fujita H, Hara T, Zhou X et al (2019) Function integrated diagnostic assistance based on multidisciplinary computational anatomy models -Progress Overview FY2014-FY2018-. In: Proceedings of the Fifth International Symposium on the Project "Multidisciplinary Computational Anatomy", pp 115–128
12. Tanimura K, Sato S, Fuseya Y et al (2016) Quantitative assessment of erector spinae muscles in patients with chronic obstructive pulmonary disease. Novel chest computed tomography-derived index for prognosis. Ann Am Thorac Soc 13(3): 334–341
13. Kamiya N, Li J, Kume M et al (2018) Fully automatic segmentation of paraspinal muscles from 3D torso CT images via multi-scale iterative random forest classifications. Int J Comput Assist Radiol Surg 13(11):1697–1706. https://doi.org/10.1007/s11548-018-1852-1
14. Kume M, Kamiya N, Zhou X et al (2017) Automated recognition of the erector spinae muscle based on deep CNN at the level of the twelfth thoracic vertebrae in torso CT images. In: Proceedings of the 36th JAMIT annual meeting
15. Zhou X, Takayama R, Wang S et al (2017) Deep learning of the sectional appearances of 3D CT images for anatomical structure segmentation based on an FCN voting method. Med Phys 44(10):5221–5233. https://doi.org/10.1002/mp.12480
16. Kamiya N, Kume M, Zheng G et al (2019) Automated recognition of erector spinae muscles and their skeletal attachment region via deep learning in torso CT images. Comput Methods Clin Appl Musculoskelet Imaging:1–10. https://doi.org/10.1007/978-3-030-11166-3_1
17. Zhou X, Ito T, Takayama R et al (2016) Three-dimensional CT image segmentation by combining 2D fully convolutional network with 3D majority voting. In: Proceedings of the Workshop on the 2nd Deep Learning in Medical Image Analysis (DLMIA) in MICCAI 2016, LNCS 10008, pp 111–120
18. Kamiya N, Ieda K, Zhou X et al (2017) Automated analysis of whole skeletal muscle for muscular atrophy detection of ALS in whole-body CT images: preliminary study. In: Proceedings of the SPIE Medical Imaging 2017, Computer-Aided Diagnosis, 10134, 1013442-1-1013442-6. https://doi.org/10.1117/12.2251584
19. Kamiya N, Oshima A, Asano E et al (2019) Initial study on the classification of amyotrophic diseases using texture analysis and deep learning in whole-body CT images. In: Proceedings of the SPIE 11050, International Forum on Medical Imaging in Asia 2019, 110500X. https://doi.org/10.1117/12.2518199
20. Oshima A, Kamiya N, Zhou X et al (2019) Automated segmentation of surface muscle in whole-body CT images using 2D U-Net: preliminary study. In: Proceedings of the IEEE EMBC2019, ThPOS-32.34, p 71
21. Wakamatsu Y, Kamiya N, Zhou X et al (2019) Bone segmentation in whole-body CT images using 2D U-Net. In: Proceedings of the IEEE EMBC2019, ThPOS-32.35, p 72
22. Klein A, Warszawski J, Hillengaß J et al (2019) Automatic bone segmentation in whole-body CT images. Int J Comput Assist Radiol Surg 14(1):21–29
23. Kume M, Kamiya N, Zhou X et al (2019) Development of representation method of muscle running using attachment region of the spinal column erector muscle in the torso CT images. IEICE Tech Rep 118(412):39–40

Index

G. Lee, H. Fujita (eds.), *Deep Learning in Medical Image Analysis*,
Advances in Experimental Medicine and Biology 1213,
https://doi.org/10.1007/978-3-030-33128-3

The manufacturer's authorised representative in the EU is Springer Nature Customer Service Centre GmbH, Europaplatz 3, 69115 Heidelberg, Germany. If you have any concerns regarding our products, please contact ProductSafety@springernature.com

Printed and bound by CPI Group (UK) Ltd, Croydon, CR0 4YY
22/09/2026
02221057-0019